Discovery of Oyu Tolgoi

Discovery of Oyu Tolgoi

A Case Study of Mineral and Geological Exploration

Sergei Diakov
Group Discovery Advisor, Anglo American, United States

Samand Sanjdorj
Emeritus Vice President of the Oyu Tolgoi Company, Mongolia

Galsan Jamsrandorj
Consultant, Exploration Geologist, Mongolia

ELSEVIER

Elsevier
Radarweg 29, PO Box 211, 1000 AE Amsterdam, Netherlands
The Boulevard, Langford Lane, Kidlington, Oxford OX5 1GB, United Kingdom
50 Hampshire Street, 5th Floor, Cambridge, MA 02139, United States

Library of Congress Cataloging-in-Publication Data
A catalog record for this book is available from the Library of Congress

British Library Cataloguing-in-Publication Data
A catalogue record for this book is available from the British Library

ISBN: 978-0-12-816089-3

For information on all Elsevier publications visit our website at
https://www.elsevier.com/books-and-journals

Publisher: Candice Janco
Acquisition Editor: Amy Shapiro
Editorial Project Manager: Ruby Smith
Production Project Manager: Nilesh Kumar Shah
Cover Designer: Matthew Limbert

Typeset by TNQ Technologies

Contents

Preamble

The discovery of Oyu Tolgoi, Mongolia, one of the largest copper–gold deposits in the world, became one of the most remarkable events in the mining industry of Mongolia during the 21st century.

The authors of the book took a direct part in the events that consequently led to the discovery of Oyu Tolgoi, which is now rightly included in the global list of unique mineral deposits. In the following pages, the authors convey their discovery story interestingly and honestly.

We are proud that both Mongolian geologists and their overseas counterparts successfully managed to combine their knowledge and experience, determination and persistence, capital and persuasiveness, which altogether consequently lead them to their final discovery goal.

Readers will get a glimpse of how Mongolian geologists worked together with their foreign colleagues. Undoubtedly, the path toward the discovery of Oyu Tolgoi, with its geological complexity, required implementation of new technologies, some of which were applied for exploration in Mongolia for the first time. For example, advanced remote sensing and geophysical methods, well covered in the book, and also technologies of efficient deep drilling to depths of 1,300 to 1,500 m and developing mine tunnels for extended distances to 2,600 m used on later stages of exploration and development of Oyu Tolgoi.

At the end of the 20th century, we hailed the completion of Erdenet Complex construction, the flagship of our mining industry. We worked together with the Soviet specialists side by side. To this day, we feel the effects of this cooperation. Former General Director of Erdenet Shagdariin Otgonbileg invited foreign experts to join their efforts in search of large copper deposits in Mongolia. For this purpose, he facilitated the creation of the joint venture Erdenet–Magma with the American mining company Magma Copper. The book lays out the details of how this process started and evolved.

The copper deposit received the name Oyu Tolgoi for an apparent reason—in the past, Mongols used this term for copper oxides on the surface. Hence the discoverers selected the name for their discovery quite appropriately.

Foreign investors became interested in the mineral potential of Mongolia for several reasons firstly because Mongolia became openly democratic, and secondly, due to the adoption of new laws favorable to foreign investments.

There were many obstacles on the path toward the discovery of Oyu Tolgoi, often driven by economic and political risks. These and other dangers could stop exploration and prospecting work in the country at any moment, as depicted by the authors in this book. Awkward words by Robert Friedland at the 2005 Investment Conference in Miami caused big protests in Mongolia, resulting in changes to the legislature. Consequently, despite our frequent expressions of dissatisfaction with R. Friedland, thanks to his effort and persistence the Oyu Tolgoi project became a success.

The leading mining company BHP, which played a pivotal role in the original discovery of Oyu Tolgoi, came to Mongolia twice. On one hand, it was related to the global economic crisis, and on the other, it was driven by the unstable mineral legislature of Mongolia. Another contributing factor was ongoing internal change and transformation within BHP at that time.

The book contains useful, educative information about prudent approaches for the resolution of critical situations in the investment policy. Maintaining a stable environment plays a crucial role in long-term investments, and the legislature must consider any changes very thoughtfully and exceptionally carefully.

The book also discusses the challenges and complexities of geological exploration and the discovery of a new mineral deposit that is economically viable. In many respects, this is more complex than extracting already discovered resources from the ground. Finding a new deposit involves a lot of unconventional thinking, persevering through various complications and challenges to determine the most effective exploration methods for different types of deposits. It is also essential to establish the right productive and collaborative environment among numerous experts on the exploration team. All these efforts may become in vain and will fail if the right setting is not established. In the book, the authors demonstrate that until Oyu Tolgoi became an obviously economically viable project and truly a world-class discovery, it was important to have a hard coordinated effort from all geologists, geochemists, geophysicists, and mining engineers.

I do not remember a previous time in Mongolia when such eloquent language could describe a devoted hard work of geologists from sunrise to sunset that they were short of time not only to show off their findings but also to explain properly the importance of their discovery to those who criticize them so often and so quickly.

Logically, for geologists, a rock is a source of information and understanding the importance of their future discovery. By nature, geologists are optimistic thinkers who have deep faith in their findings. Out of almost 8,000 mineral prospects registered in Mongolia, few will become commercial deposits. Discovering such economic deposits requires enormous energy and effort. Every so often, when such a successful discovery occurs, it makes a tremendous positive impact on the country's economy.

First President of Mongolia, Doctor of Science.

Punsalmaagiin Ochirbat

Discovery requires a complex sequence of actions and events based on facts, interpretation of those facts, guesswork, skill, and circumstances beyond one's control coupled with a proper dosage of serendipity and perseverance. Discovery leading to the identification, development, and eventually construction of a significant metal mine in a remote part of the world lacking modern infrastructure furthermore requires a sustained devotion to a belief in success as well as corporate financial support along with good governmental relations.

None of these actions is easy at the time, nor are they always made correctly, as the history of many deposits and prospects tells us. Understanding these events is essential for any exploration geologist, but rarely the details of the process leading to success are making to the pages of publications. Fortunately, it is possible to learn from the past, if one takes the time and spends the effort. The book entitled "*Discovery of Oyu Tolgoi: Case Study of Mineral and Geological Exploration*" by S. Diakov, S. Sanjdorj, and G. Jamsrandorj fills that gap in knowledge. In the book they describe in chronologic order the sequence of events and decisions that over 20 years culminated in the discovery of one of the major copper and gold resources found in the last 50 years.

Beginning with a brief history of Magma Copper before the entry into Mongolia, the book follows the sequence of events through the decision to explore in Asia along with various governmental and corporate decisions required to embark on what is well known to be a long and tortuous path to a significant mineral deposit. Critical to the successful process outlined is the crucial and sometimes fortuitous steps that aided the rapidly changing political environment in Mongolia and facilitated a desire of a mostly agrarian country to begin to diversify and transform their economy. For many developing countries, this requires the usage of their natural resources and a set of legal and environmental regulations that provide a solid basis for economic activity.

The book outlines the corporate and governmental steps and missteps that were made along the path that resulted in the identification of the porphyry Cu–Au deposits now being mined in the Gobi Desert of southern Mongolia. Advancement toward the discovery success started with the initial decision of Magma Copper to explore in Mongolia through the establishment of the joint venture exploration agreement with the Erdenet Mining Company, the operators of the Soviet-constructed porphyry Cu–Mo mine in northern Mongolia. Before moving out in the field, the joint venture conducted a compilation of all geologic and mineral occurrence data and took decisions on which types of metal commodities to direct their focus on, followed by wide-ranging reconnaissance field reviews guided by the data compilation. The result of these initial steps in the exploration program led the explorers to focus their attention on porphyry copper targets in the accreted magmatic arcs in the Gobi, as these types of deposits were known in the region at the time. Furthermore, they had the potential to be of significant economic value to support the development of the required infrastructure in an area without roads, power, or even towns. After closing the joint venture, BHP decided to continue copper exploration in Mongolia on their own. This sequence of events finally brought them to the Oyu Tolgoi area with the resulting discovery and recognition of outcropping porphyry copper style veins in a couple of small hills rising a few meters above the desert terrain.

Accompanying or perhaps overriding the on-the-ground geologic, logistical, and political events, as well as corporate decisions, personnel changes at all levels, takeovers, and mergers, and ultimately changes in exploration strategies intertwined with the internal corporate approach, are presented in a factual manner. This gives the readers an opportunity to follow the dynamics of these events during the early stages of the discovery of Oyu Tolgoi deposit. In the end, it was a corporate decision based on their internal economic analysis that paved the path for a risk-adverse project owner to exit the project to be replaced by the risk-taking company. The last change in managers of the exploration program provided the final impetus to identify the full mineral potential of the prospect and to the identification of the vast quantity of metal that has become the known Oyu Tolgoi mine.

As stated in the introductory pages, the book is not intended to be a scientific document, although historical, geologic, and geophysical maps and other observations critical to decisions are described. Instead, the book is probably one of a few books that dispassionately outlines the history of the discovery of a significant metal mine, giving credit to all who contributed but also not neglecting those that may have unwittingly hindered the process. In this context the book serves a vital part of the historical geology lore, and all young geoscientist entering the business of resource exploration and extraction should take the time to become aware of the lessons presented.

Richard M. Tosdal, Ph.D., P.Geo
Folly Beach, South Carolina, U.S.A.

List of Abbreviations

ADEQ Arizona Department of Environmental Quality.

ALS Australian Laboratory Services, a laboratory service testing provider, currently one of the world's largest providers of laboratory analysis services to the minerals industry.

BLEG Bulk Leach Extractable Gold, a geochemical method of sampling and analysis used while prospecting for gold. The technique was developed in the early 80s for accurate measurement of fine gold in bulk samples from 2 to 5 kg using cyanide leaching extraction technique.

BHP Broken Hill Proprietary, a major mining company with headquarters in Melbourne, Australia. BHP Minerals and BHP Copper were production divisions of BHP. After the merger with Billiton became BHP Billiton. Currently returned to its old historical name BHP.

C1, C2 categories of the Russian system of a mineral resource.

CEO Chief Executive Officer.

CMEA Council for Mutual Economic Assistance (COMECON or CAME), an economic organization from 1949 to 1991 under the leadership of the Soviet Union that comprised the countries of the Eastern Bloc.

cm centimeter.

DC diamond core drilling.

E east.

FS feasibility study.

FOB free on board, shipping rates.

GAZ Gorky Automobile Plan, vehicle brand produced in Russian Federation.

GDP gross domestic product.

HBI hot briquetted iron.

ICP induced coupled plasma optical emission spectrometry for detecting and analyzing trace elements.

IOCG iron oxide copper–gold deposits.

IP geophysical method of induced polarization.

JICA Japan International Cooperation Agency.

JORC Joint Ore Reserve Committee—a system of classification of mineral resources/reserves.

JV joint venture.

K–Ar potassic-argon.

LIBOR London Interbank Offered Rate, average percent for loans from London banks.

LLC Limited Liability Company, or Ltd.

Ma million years ago—geological age in million years.

mm millimeter.

MRAM Mineral Resource Agency of Mongolia.

N north.

NE northeast.

NI 43-101 National Instrument for the Standards of Disclosure for Mineral Projects in Canada.
NW northwest.
OT Oyu Tolgoi
PDAC Prospectors and Developers Association of Canada, union of prospectors and explorers of Canada.
PIMA Portable Infrared Mineral Analyzer, equipment to measure mineral spectrum in the field.
PhD Philosophy Doctor, scientific degree.
ppm parts per million.
RC reverse circulation drilling.
RGB Landsat images in red-green-blue color balancing spectrum.
SBA Soviet made drilling rig.
SEG Society of Economic Geologists, a global union of economic geologists studying mineral deposits.
SPOT French Satellite Pour l'Observation de la Terre or English: Satellite for observation of Earth, satellite images with high resolution. The system originated in France and then supported by European Space Agency in the 70 and 90s.
SE southeast.
SW southwest.
SXEW Solvent Extraction and Electro Winning, a method of metal extraction from solutions, usually for copper.
TEM Transient Electro Magnetic, a geophysical method based on principals of electromagnetism.
TM Thematic Mapper, a variety of Landsat satellite images.
TMI Total Magnetic Intensity
UAZ Ural Automobile Plant, vehicle brand produced in Russian Federation.
UK United Kingdom.
USA or US United States of America.
USGS Geological Survey of the United States of America.
USSR Union of Soviet Socialist Republics.
UTM Universal Transverse Mercator coordination system.
VMS Volcanic Massive Sulfide deposits.
VES Vertical Electric Sounding, the geophysical method providing information about geological structure based on electrical conductivity or resistivity of the medium.
WMC Western Mining Corporation, WMC Resources Ltd. was acquired by BHP Billiton in 2005.
ZIF Soviet made drilling rig.

Mineral abbreviations used in the text:

ars arsenopyrite
az azurite
bn bornite
cc chalcocite
chp chrysoprase
chr chrysocolla
cp chalcopyrite
cv covellite
cu cuprite
ga galena

hm hematite
mag magnetite
mal malachite
mol molybdenite
nat native copper
py pyrite
pyr pyrrhotite
sp sphalerite
tn tenorite.

Introduction

Dedicated to our families

The discovery of the Oyu Tolgoi copper–gold deposit, made almost 20 years ago in the Southern Gobi Desert, was a remarkable event in the history of modern Mongolia. As one of the most significant global copper–gold porphyry discovery in recent years, this discovery marks a significant milestone in the history of Mongolian mining industry and the copper industry worldwide.

The discovery of this deposit was the culmination of 2 years of exploration activity first performed by the Erdenet–Magma Joint Venture and subsequently by BHP geoscientists between 1995 and 1997.

It appears that the early stages of this discovery were veiled in a shroud of mystery, which has led to much erroneous conjecture and speculation over the years. The authors of this book, all of whom participated in the development and implementation of the exploration program, believe that the earliest stages of geological exploration are probably the most critical and often the most challenging in the process of any mineral discovery. In this book, they present a step-by-step detailed account of the sequence of events that led them to this world-class discovery and share their thrilling experiences from the early days of the deposit's discovery history.

A critically important aspect of any geological exploration program is to conduct adequate research during the "desktop stage" of the program to be able to select, with a reasonable degree of confidence, the most geologically permissive exploration terrane. Research may include some preliminary field reconnaissance. Failure to locate a fertile terrane at an early stage will generate discouraging field exploration results, which will delay or prevent the realization of the final goal of success. Also, methods and approaches selected early have a significant impact on the future of the exploration project. There are many examples when early stage exploration, despite the successful identification of a terrane with favorable geological conditions, was doomed to failure due to various factors.

Our success at Oyu Tolgoi did not come instantly. The path to success was thorny. We describe the obstacles in our way and how we managed to surmount them. We also convey the story of the western mining companies' growing interest in the copper potential of Mongolia during the late 1990s and how the Oyu Tolgoi discovery in the South Gobi contributed to this. We are convinced that this historical account will be of significant value to the reader.

Of course, during the research or desktop study phase of the program, we made use of the valuable data produced by the previous generations of explorers who conducted regional surveys and geological mapping and pursued their exploration goals in the past. Free access to geological data from The Geological Funds of Mongolia played a critical role stimulating the growth of exploration interest in the country. We acknowledge the effort of all geologists and mineral experts for their dedicated work and their contribution to the development of the mineral resource base of Mongolia. Their hard work and achievement also contributed to our success.

We believe that publication of the details of the anthology of this discovery will provide some insight into the vagaries of the thinking process resulting in the discovery of the most extensive copper porphyry deposit in Mongolia. Without prejudice, we attempted to shed some light on the discoverers' thought process, what decisions we made, and why we made them. Knowledge and understanding of how the discovery process started will undoubtedly be instructive to them. Hopefully, the lessons conveyed in this book will be particularly useful to the young geological explorers especially those who are dreaming about their future discoveries.

In the following pages, we also touch upon the makeup and interactions of the diverse exploration team, their psychology, mentality, and how corporate cultures from previous employment affected their decisions. In many ways, the culture has deep roots in the past, playing a pivotal role in everyone's behavior and attitude toward business. We adopted new approaches in our thinking. In our view, this played a significant role in our success. At first glance, it appeared that a group as diverse and multicultural as ours could not work productively. On the contrary, against all the odds and skepticism, the team in its determination to deliver a significant discovery was surprisingly effective. A combination of knowledge and wisdom from veterans and ambitions and aspirations of young geologists worked quite well for us. Neither age nor language barriers stopped productive communication. We are not claiming that everything happened smoothly. There were challenges. The details of unfolding events of how we managed these various problems on our path to success are the story inscribed on the pages of this book.

The authors present the events leading to the discovery of Oyu Tolgoi in chronological order. We portray these facts as we see them after more than 15 years since they took place. We have no ambition to misrepresent the truth, nor do we intend to claim to have a magic recipe for success. This book is not a technical report, nor is it a scientific publication. The use of maps, cross sections, and diagrams, along with other technical information, helps us to convey our story more clearly.

We hope that while reading this book, the readers will find useful insights and ideas for their own practical needs. We believe that in the end, this story will facilitate them in their strive for success and consequently will become a useful guide for them toward their discoveries.

During the lifetime of every economic mineral deposit, there is a period of transition from exploration to production. Ultimately, mineral discoveries are made for the benefit of future metal production. This all-important transitional stage, which can make or break the economic viability of a deposit, has many challenges. At Oyu Tolgoi, these problems are currently unfolding in front of our eyes. In 2009 Ivanhoe completed construction of the mining and processing complex at Oyu Tolgoi, and Rio Tinto started production in 2010 by expanding the producing complex to its optimum capacity. It may now be timely to reflect on the past and review those initial moments that catapulted Oyu Tolgoi on the path to success.

Chapter 1

Prologue: Foundation for Future Success

Chapter Outline

Burgess J. Winter *was born in Ireland. He graduated from College of Technology in Belfast in chemistry. Burgess started his professional career in the copper belt of Rhodesia in 1959, and since 1973, he continued with Anglo American. He managed copper operations in Swaziland, Brazil, and America. In 1976 Burgess came to the Unites States of America to lead the division of BP Minerals (previously Kennecott) as Senior Vice President. In 1988 he became the President and Director of Magma Copper Company. In 1994 Burgess received a prestigious mining award, and he was inducted into the Hall of Mining Fame. In December 1996 he retired from BHP.*

Discovery of Oyu Tolgoi. https://doi.org/10.1016/B978-0-12-816089-3.00001-9

Bradford A. Mills *received Master's degree in Geology and Mineral Economics when he graduated from Stanford University in 1979. Brad is currently the Executive Chairman of Mandalay Resources, based in London, UK. Brad is also a Director of Plinian Capital and recently was on the Board of Directors of Norilsk Nickel. Previously, he was the CEO of Lonmin Platinum and President of Base Metals of BHP Billiton and Vice President of Strategy. In Magma he occupied a position of Executive Vice President and Vice President of Business Development.*

Shagdariin Otgonbileg *graduated from Irkutsk Polytechnic Institute in 1977 majoring in mining survey. He was appointed to the position of General Director of the Erdenet Mining Complex. Otgonbileg made a significant contribution in bringing Erdenet to the international level by establishing supplying lines of Erdenet product to dozens of countries. He initiated a concept of technological improvement, including processing of copper oxide ore—the foundation for production at Erdenet copper cathode through ErdMin enterprise. In 2001, Otgonbileg was tragically killed in a helicopter crash in Western Mongolia while reviewing the area of natural disaster with the UN delegation.*

1.1 MAGMA COPPER COMPANY STORY: WHY AND HOW MAGMA CAME TO MONGOLIA?

The Magma Copper Company, or Magma headquartered in Tucson, Arizona, was one of the significant copper producers in America during the 1990s with copper mines in Arizona and Nevada. The flagships of the company's copper production were San Manuel, Kalamazoo, Pinto Valley, the Magma mine at Superior in Arizona, and the Robinson copper-gold mine in Nevada. Additional copper came from the Tintaya mine in Peru. The company owned a smelter located in a small mining town in San Manuel, northeast of Tucson. Magma produced copper cathodes and copper wire. At that time, there were discussions about the advanced processing of copper in the industry such as the production of copper powder, allowing manufacturing of various copper configurations by pressing the powder, which promised to deliver high returns.

Magma has a long history beginning in the 19th century. It started with the Silver Queen mine in 1890 initially producing silver. However, later, after acquisition by William Boyce Thompson, the production switched from silver to copper mined from rich copper ores hosted by series of subparallel veins. In 1910 Bill Thompson registered a new company named Magma Copper Company. In the 1950s Magma became the eighth largest copper producer in the United States. In 1956 the company initiated production from its San Manuel mine and then claimed the neighboring area of Kalamazoo.

In 1969 Newmont Mining Corporation (Newmont) absorbed Magma Copper. In 1987 Magma again became an independent company after Newmont in their effort to focus purely on gold, deciding to spin off its copper-producing division. However, Newmont maintained 15% of its stock in Magma. Building a mine operation and smelter plant in San Manuel compliant with newly established environmental standards and regulations kept Magma in the "red zone" as far as profit was concerned for the following 11-year period until 1988 when the copper price went up from US$0.60 to US$1.50 per pound. Magma built a new furnace at San Manuel making the smelter (Fig. 1.1) one of the largest in the world with a 21% total smelting capacity in the United States of America.

After the sudden death of Magma president Brian Wolfe, the reins of the company transferred to Burgess Winter. In 1989, Magma managed to buy out the remaining portion of the 15% stock from Newmont, becoming a full owner

FIGURE 1.1 Magma Copper smelter in San Manuel (Briggs, 2013).

of its assets. In 1990, the company achieved some spectacular results, including increasing its annual copper production significantly to 213,100 metric tonnes. From this output, 52,000 tonnes of copper came from acid-leach operations, which was in fact 24% higher than that in the preceding year.

Magma's San Manuel copper mine was located northeast of Tucson in the Pinal County at a distance of about 65 km. The mining activities began in the year of 1948. Initially, mining at the San Manuel porphyry copper ore body started as an open pit, and later on, the operation switched to underground mining. After the discovery of the displaced part of the ore body, named Kalamazoo, the mine became the most significant underground mining operation in the United States, delivering to the surface about 20 million tonnes per year through several vertical shafts. During the mine history (mining was suspended June 25, 1999), Magma extracted and brought to the surface a total of 702.9 million tonnes of material. It includes 17.1 million tonnes of country rocks and 624.9 million tonnes of ore from the San Manuel operation and 50.4 million tonnes of ore from the Kalamazoo deposit (ADEQ, 2013).

A heap leaching operation method was used to treat the oxidized ore from San Manuel's open pit. The mined oxidized ores were stacked in piles near the open pit. Copper after dissolution by acid solutions was transported to the tanks where it was extracted from the solutions by a method of electrowinning. The remaining copper in the oxidized ores in the slopes of the open pit was also removed from the depths, but in a slightly different manner, which is by in-situ leaching with subsequent production of copper cathode similarly, through electrowinning process.

Geologically the mine area is composed of a Cambrian rock complex (monzonites, granodiorites, diabase, and aplites), overlain by Pliocene-Pleistocene conglomerates of Gila suit as well as Quaternary conglomerates of the nearby River San Pedro (Force et al., 1995). Mineralization is associated with the body of granodiorite porphyry of cylindrical shape with dimensions of 2,650 m in length and 830-m wide and situated at a depth of 230 to 1,000 m below the surface. The intrusion of granodiorite porphyry mineralization, judging by the current position of Gila conglomerates, was tilted on its side and then dissected by a fault, along which the top half of the ore body called Kalamazoo was dropped over a distance of about 2.5 km. Initially, only a part of the body located in the footwall of the tectonic fault was found. From this section, the San Manuel mine started. Later on in the early 1960s, after the data compilation by David Lowell, it was suggested that the known part of the ore body was only a half of the originally mineralized intrusion. Therefore a displaced part must be somewhere at depth. Subsequently, David Lowell in his article (Lowell, 1968) outlined the lateral and vertical symmetric zoning of porphyry deposits by combining the San Manuel and Kalamazoo ore bodies into one single ore body. A joint article with the professor of the University of Arizona John Gilbert applied this model to assess porphyry deposits known in the North American Cordillera (Lowell and Guilbert, 1970). The impetus for such a model was an attempt to explain the asymmetrical zoning type of mineralization and alteration observed at San Manuel, which well supplemented the discovered part of mineralization within the Kalamazoo ore body. After confirmation of the displaced portion of the ore body at depth, it became clear to all that the model had important practical significance for further prospecting and exploration of porphyry deposits. Of course, subsequently, the model was improved showing that zoning is not always necessarily symmetrical and that zoning could have a more complicated shape. Nevertheless, at that time it was a significant breakthrough in the understanding of copper porphyry systems.

Mineralization of the San Manuel ore body consisted of primary sulfide chalcopyrite mineralization and secondary copper oxides; among oxide copper minerals, chrysocolla was dominant. The average content of copper in the primary ores was 0.65% and molybdenum at about 0.015%. The ore body in Kalamazoo after drilling six exploratory drill holes was intercepted at depth from 800 to 1,300 m (Sandbak and Alexander, 1995). These ores required traditional enrichment with the copper concentrate production and then subsequent metallurgical processing through smelting at the company plant located in nearby town of San Manuel. Oxide ore went through a much more straightforward process of treatment for copper production through either heap or underground leaching.

In 1991, Magma acquired copper–gold mines in the Robinson mining district in Nevada near the small, isolated mining town of Ely. A feasibility study indicated that the optimum capacity of the mines in the district would be an annual production of 56,700 metric tonnes of copper and 86,500 ounces of gold if production continued uninterrupted for a duration period of 16 years. Management of Magma and its labor unions had to agree principally and decisively on stability terms and conditions. By October 1991, after numerous discussions and heated debates, they came to a new labor agreement, which at that time was unprecedented in the history of labor relations in America. The parties concurred on a contract, which warranted stable, uninterrupted production with no labor strikes for 8 years. The agreement stipulated that representatives of the company management and trade unions would supervise the company operation jointly.

This labor agreement put Magma in a favorable situation compared with other mining companies in the United States of America. The working relations between company management and labor union leaders were strained, if not on "a war footing." A labor dispute at Phelps Dodge, another copper producer in Arizona, resulted in a fierce standoff from 1983 to 1986, which nearly ended in a bankruptcy situation.

The approach by Magma management team was different. They decided to abandon the standard centralized approach in which management made all the crucial decisions and switched the lines of communication from vertical to lateral. After that, employees discussed all critically essential issues collectively with their managers, reached collaborative decisions, and implemented them as a team. With this new management structure, all personnel from managers, supervisors, union leaders, to workers of the lowest levels took ownership of these decisions. Consequently, it became clear that the atmosphere of inclusive, sincere discussions was much more productive.

At first glance, such collective approach may seem counterproductive. Indeed, it required lengthy discussions with numerous personnel with different views and opinions on how the production line should run. However, this approach raised responsibilities of all employees in the company from miners to metallurgists, from engineers to geologists.

The assets that Magma inherited from Newmont were plagued with numerous problems. Not only did Magma receive Newmont's nonperforming assets but they were also burdened with complicated labor relations. In Burgess' opinion the fortunes of a company with inferior hard assets could be enhanced with excellent labor relation practices. The improved working environment would in all likelihood result in better and less stressful working conditions, leading to increased productivity, which would be beneficial to all. Magma management team saw the path toward improved efficiency through complete trust in the ability and decisions of its employees. Workers, despite their diverse backgrounds and different viewpoints, pulled together as a team striving for improved production and efficiency. As a result, there was complete trust along the whole chain of command. Vertical lines turned into horizontal, and people started performing much better as a team of common thinkers and contributors rather than direct top-down line reports. Another outcome was that management could devote more time to finding solutions to issues rather than supervising personnel.

Interestingly, some of the workers started putting many of the lessons learned in the workplace into practice in their homes. Magma management recognized the importance of maintaining a right balance among work schedules, family relations, and employees' rest and recreation. Finding an effective solution for domestic issues, harmonizing personal relations in the family, also contributed to the improvement of workers' productivity.

Removing the barriers inherent of the old centralized system elevated productivity, resulting in a positive impact on the company's performance and the lifestyle of the employees.

Burgess believed that the company's miners could also be value creators, much like their highly educated supervising mining engineers. The consensus of Magma's leadership was that the labor force would be strongly motivated if the barriers to their involvement in business management were entirely removed allowing them to participate in the process of growing production. The Landmark Forum was selected to cement this relationship.

The quintessential part of the Landmark Forum focuses on the following. There is a big difference between what happens in everyone's life and how these events are interpreted. People, in general, pursue the achievement of self-satisfaction in their lives. The desire to look good drives human behavior.

People explain the events in their lives, and these interpretations do not necessarily coincide with the reality and truth. In striving toward success, people complain about difficulties in their lives, and this is the least productive approach to the solutions. Individuals, instead of reinventing themselves, find new ways for effective solutions by introspection. At the core of the Landmark's philosophy is that the self-introspection can open up new possibilities. Among the people who took the course, the opinions range from extremely negative to extremely positive. The critical point in this process is an opportunity to look at one's self, which helps to find the most rational solution to their problems or challenges both at work and at home.

Victor Forberger (2002) described this innovative approach by Magma, which allowed it to reach success in negotiations with the labor unions and then consequently in the development of new working relationships. The Magma workers started creating their challenging goals and began achieving those goals. They did not require many directions from the top. The result was an enthusiastic search for the most efficient solutions in the company, whether in the underground or surface mining operations, or at the company mill or its smelter or in the enterprise office, everyone started looking for them across the organization. The after effect of this was a significant increase in discipline, production capacity, improved safety, and better work environment. Magma transformed into a profitable prosperous company to become an employee of which was a big fortune and great luck.

Magma's Bradford A. Mills was responsible for business development from 1989 to 1994. Brad facilitated extraordinary thinking and encouraged the search for practical solutions for the most challenging issues faced by Magma personnel at various levels.

One of the examples of the successful outcome for the extremely challenging situation was the discovery of Magma Porphyry at Superior. In 1995 Magma stopped its copper production from the underground mine (Fig. 1.2)

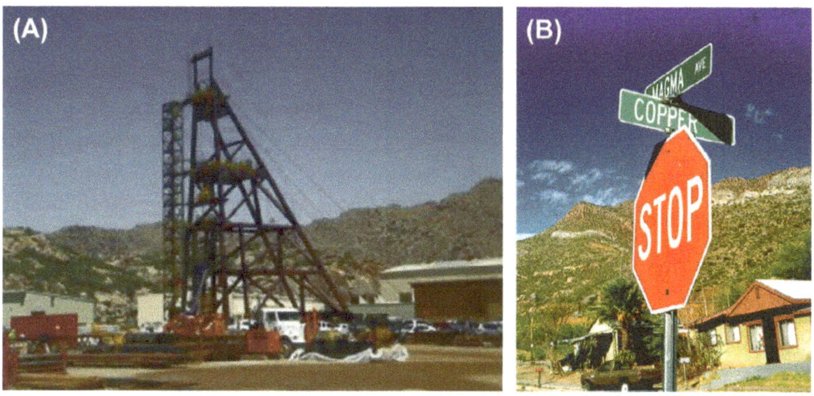

FIGURE 1.2 Tower shaft #9 of the Magma mine (A) and a street junction in the mining town of Superior, Arizona (B), photos 1996.

due to depletion of reserves. Under the guidance of Senior Project Geologist Alex Paul and geologist Scott Manske, a concerted effort was undertaken to expand the resource/reserve base using the Magma parallel vein model. This model proved to be unsuccessful. Nevertheless, while looking for new parallel veins, observations made by the geologists led to the discovery of a different style of mineralization with the more significant resource.

During this time, Magma also initiated its Mongolia exploration programs. In the early stages of both the Superior drilling and the Mongolian efforts, it became apparent that Magma's traditional copper exploration approach would have to be modified to accommodate realistic geological circumstances.

At the Superior mine, the surface geology in the outcrops is postmineral dacite tuffs of Miocene age called Apache Leap with Oligocene–Miocene conglomerates of the Whitetail Suite volcanics at the base. They are overlying Mesozoic sedimentary, intermediate volcanic and volcaniclastic rocks overlying Paleozoic limestones of Pennsylvanian–Permian Naco, Mississippian Escabrosa, Devonian Martin, and Cambrian Bolsa quartzite formations. Precambrian Apache Group with Dripping Springs quartzites were at the bottom over early Proterozoic age Pinal Schist formation. Thick sills of Cambrian diabase cross cut Precambrian rocks. These are the host rocks for steeply dipping Magma veins hosting copper mineralization. Mining production came from ore veins carrying quartz–pyrite–chalcocite–bornite mineralization extending for a vertical distance of about 1,500 m and up to 600 m in strike length varying from 2 to 10 m in thickness with average grades of 4.5% copper, 53 g/t of silver, and 0.91 g/t of gold. This vein contained more than 10 million tonnes of ore (Paul and Knight, 1995). The exploration drilling was performed from the low mine horizons horizontally across a strike of the copper veins.

Discovery of another mineralized vein could extend the life of the Superior mine for another decade or so. However, exploration drilling intercepted hydrothermal alteration with weak quartz–sericite–pyrite assemblage at 1,300 m below the surface. The geologists interpreted this to be indicative of an underlying porphyry system, which, if mineralized, would probably generate the high-grade Magma veins at the upper levels. To test this hypothesis, the exploration strategy was changed from drilling subhorizontal drill holes to drilling inclined holes to explore for a copper porphyry system.

A five-drill-hole program (two from the surface and three from underground) intersected higher than 2% copper in four of the holes, the deepest being 1,950 m and indicated the presence of a significant porphyry copper mineralization grading from 1% to 2% copper (Paul and Manske, 2005). It was named Magma Porphyry (Manske and Paul, 2002).

Based on the dip angle of the postmineral Apache Leap volcanic tuffs, the Magma Porphyry tilted to the east at an angle of 25 degrees (Paul and Manske, 2005). Felsic porphyry dikes of Laramide age intruded the

Proterozoic–Paleozoic–Mesozoic rocks, which host the Magma Porphyry mineralization. On the periphery, drilling intercepted a typical propylitic alteration carrying quartz veins with pyrite–specularite in association with disseminated sphalerite and chalcopyrite. The central part of the porphyry system carried quartz–pyrite–sericitic alteration also hosting disseminated pyrite, bornite, and chalcopyrite. Below this, there was a zone of potassic alteration with numerous flakes of biotite. Potassic alteration superimposed over sericitic alteration. Upper Paleozoic carbonate rock of Nacho Formation contained zones of hornfels and skarns. Drilling at depth intercepted stocks and dikes of porphyritic rocks, which most likely brought hydrothermal solutions with mineralization. Their age ranged from 66 to 69 Ma (Paul and Manske, 2005).

Initially, the zone of mineralization higher than 1% copper had 750 m in strike length, 250 m in width, and 300 m in depth was estimated. Besides copper, the mineralization also carried molybdenum at an average grade of 0.02%. The upper part of the mineralized body was at a depth of 1,500 m from the surface (Paul and Manske, 2005). Although BHP stopped the exploration work, it was clear that Magma Porphyry was a new significant porphyry discovery in North America.

Analogous to Grasberg deposit, Magma Porphyry confirmed that to form a high-grade economic ore body, it is not necessary to have a secondary enrichment like in many other porphyries in Southwest United States and South America. If there is a reactive host rock or another grade-enhancing factor present, then hypogene porphyry could form an elevated tenor of copper mineralization without classic supergene enrichment. By this time, BHP intensively mined Escondida deposit in Chile via the open-pit method. The Magma Porphyry ore body at 1,500-m depth presented both logistical and technical challenges for BHP Copper management. These concerns prompted termination of production and exploration work until a suitable joint venture (JV) partner was identified to share the technical risks and associated costs.

Kennecott Exploration Inc., a daughter company of Rio Tinto, soon came forward and acquired 55% interest in the project. From 2001 to 2003, Kennecott conducted additional exploration to delineate the Magma Porphyry ore body then renamed Resolution. After their delineation drilling, the size of the ore body significantly increased to 1,600 m in strike length and 550 m down dip, enclosing the mineralized rock of 1,730 million tonnes with an average grade of 1.52% copper and 0.04% molybdenum (The Resolution Copper Project, 2015). These findings propelled Resolution to high-ranking levels of copper deposit in both North America and globally, positioning it in one line with such giant deposits as Escondida, El Teniente, Grasberg, and so forth.

As indicated earlier, Magma while advancing exploration in Superior district and conducting exploration near Tintaya in Peru also initiated exploration in Mongolia in partnership with Erdenet.

1.2 ERDENET STORY: HOW ERDENET BROUGHT MAGMA TO MONGOLIA

At the time of the Erdenet–Magma JV creation, Erdenet was the most significant copper producer in Mongolia.

The history of Erdenet began in 1963 when the Soviet field expedition lead by geologist Ushakov conducted 1:25,000 scale geological mapping in the Selenge and Orkhon River basin area within the Bulgan aimag jurisdiction. This program indicated a modest copper resource of 150 million tonnes. In the following year the geologists M. Kuzhvart, M. Krauter, and Dugersuren of the Czechoslovakian–Mongolian expedition focused their attention on the outcrops of Erdenetiin Ovoo. This effort was a part of the international cooperation among the socialist countries Council for Mutual Economic Assistance (CMEA). During the period from 1964 to 1972, the exploration expedition under the guidance of V. S. Kalinin conducted a substantial exploration program. A new resource measuring 2,500 m in strike length with a width of 500 to 1,500 m down to a depth of 400 m was delineated. The body was estimated to contain 612 million tonnes of category C2 ore with additional 670 million tonnes of combined C2 and C1 categories according to the Russian system of mineral resource classification. Mongolian geologists L. Myagmar and G. Sanduijav made significant contributions to the deposit exploration.

Construction of the mining town of Erdenet and the Mining-Beneficiation Complex of Erdenet or the Mongolian–Russian Joint Venture Enterprise LLC started in 1974 (Fig. 1.3). A railroad spur connected Erdenet with the central railroad Irkutsk–Ulaanbaatar–Beijing. By 1978, the enterprise began commercial production. Its beneficiation plant had an initial annual processing

FIGURE 1.3 Town of Erdenet, third largest city of Mongolia.

capacity of 4 million tonnes, the largest in the country. In 1980 the capacity was doubled, and by 1981, this was doubled again reaching a capacity of 16 million tonnes annually. In recent years its capacity has increased further to 22.2 million tonnes per annum, producing 126,700 tonnes of copper concentrate and 1,900 tonnes of molybdenum concentrate, which accounted for about 13.5% of the gross domestic product of Mongolia. The plant employs more than 8,000 workers.

The geology of the deposit and the region has been described in numerous publications, the main ones being studies by R.A. Khasin (1971), Marinov et al., (1977), S. Gavrilova and I. Maksimyuk (1990), O. Gerel and B. Munkhtsengel (2005), and V. Sotnikov and A. Berzina (1984). Early Cambrian metamorphosed volcanic rocks and early Paleozoic diorite and granodiorite intrusions within the Permian basalt−andesite−rhyolitic volcanic sequence of the Selenga volcanic belt host the Erdenetiin Ovoo deposit (Marinov et al., 1977). The volcanic rocks form an anticline striking northwest direction. These early Paleozoic rocks are uncomfortably overlain by a subhorizontal sequence of volcanic rocks of Triassic−Jurassic age. A subhorizontally lying sequence of volcanic rocks of Triassic−Jurassic age are uncomfortably overlying these rocks. A fault system occupying northwestern, meridional, and latitudinal directions played an important role in the localization of the intrusion and subvolcanic bodies hosting porphyry copper−molybdenum mineralization. Intrusions are of alkaline affinity ranging from quartz-monzonite and monzonite to quartz syenites in composition.

Exploration delineated two main ore bodies, the Northwestern and Central (Marinov et al., 1977). The Northwestern ore body has a strike length of 2.8 km and is 1.3 km in width. The smaller Central ore body has dimensions of 1.4 km in strike length and 0.4 km in width. The second- and third-phase granodiorite porphyry and granodiorite intrusions contain a major portion of the mineralization. A zone of oxidized mineralization extends down to a depth of 30−60 m, and a zone of secondary enrichment occurs at depths of 200−400 m. Below this, there is a sulfide zone with copper grades of 0.3%, except in intensive fracture zones with copper grades increasing to 1%−2%. Molybdenum grades vary from 0.001% to 0.76% (Marinov et al., 1977).

The highest grades of copper occurred in the upper parts of the ore bodies with an average copper grade 0.76%. The tenor of the sulfide ore was lower, at an average grade of 0.44% copper and 0.017% molybdenum. Gold and silver were insignificant averaging 0.01 and 7 g/t, respectively.

The mining of the ore bodies was by the open-pit operation (Fig. 1.4). Until 1995, the historical production of the Erdenet enterprise was 1.57 million tonnes of copper metal. At that time, the remaining resource was 5.5 million tonnes copper in the primary ore at around 0.4% copper grade.

The dominant minerals in the zone of oxidation and leaching are limonite, malachite, azurite with accessory occasionally turquoise, and chrysocolla. Chalcocite, covellite, and, to lesser extent, bornite dominate the zone of

FIGURE 1.4 Open pit of Erdenetiin Ovoo mine.

secondary enrichment. Primary mineralization is composed of pyrite, chalcopyrite, and molybdenite, occasionally magnetite, hematite, sphalerite, and tetrahedrite.

Besides Erdenetiin Ovoo, there was another known copper porphyry deposit in Mongolia, Tsagaan Suvarga, located in South Gobi. Initial resource for this deposit was 240 million tonnes averaging 0.53% copper and 0.018% molybdenum, with negligible amounts of silver and gold. Some geologists believed that Tsagaan Suvarga and Erdenetiin Ovoo may have the same submeridional regional fault (Sotnikov and Berzina, 1989). However, subsequent mapping did not support this concept, and the idea has been completely abandoned.

1.3 INVESTMENT CLIMATE AND ECONOMIC SITUATION IN MONGOLIA IN THE EARLY 1990S

We should note here that at the beginning of the 1990s, the Mongolian economy was in a status of economic shock. The CMEA that provided donor assistance to Mongolia fell apart in 1991, and subsidies from the Soviet Union were ceased. The financial injections from these sources had constituted almost one-third of the national gross domestic product. Therefore the impact was severe and devastating.

However, the country quickly realized that outside help was not available in the immediate future and that they would have to find the solutions on their own. To avoid a complete economic collapse, it was essential to initiate a complex set of reforms and reorganizations. All political forces in the country got involved in this process. Part of the transformation included political reforms changing the structure of political power. This was a significant step in transforming the economy from a centralized system to one modeled after the Western economy. Despite some political tensions such as those between the Mongolian People Revolutionary Party and the opposing Democratic Union, the politicians ultimately succeeded in creating a favorable environment for foreign investment.

Traditionally, the Mongolian economy was based on agricultural, nomadic grazing, and animal farming. Oil products, machinery, clothing, and construction materials were mostly imported. Mining played a vital role in the creation of its GDP (Wikipedia, 2015a,b,c). Production of copper, coal, fluorite, and gold were essential sources of country's hard currency income. Since 1977, a significant component of hard currency was coming from Erdenet's production of copper and molybdenum concentrate. Besides copper, there were substantial resources of coking coal, fluorite, phosphates, and other minerals, dramatically increasing the mineral resource investment potential of the country. To unlock this potential, it was necessary to make relevant economic and legal changes by the liberalization of prices, adjustments to the banking system, and recalibration of the legal system pertinent to the market economy.

The first steps were the most difficult. In 1990–92 inflation was measured by hundreds of percentage, unemployment was consistently rising, and the supply of goods and services was unpredictable. Food, for example, was rationed through a distribution system. During this challenging time, industrial production in the country dropped by one-third, and unemployment was skyrocketing. From January 1991, all trade relations with the other CMEA states switched to hard currency. Mongolia initiated tax reform and started the process of privatization. All these changes proved to be an active catalyst for conflict among different societal groups in Mongolia. It was a difficult time for the country, and it was especially difficult for the population living in the rural areas. It appeared that there would be no end to the downward spiral of the country's economy, and the people's patience was running thin.

By the mid-1990s, the situation in the country stabilized somewhat when annual growth increased to 6%. It was partly due to the changes that had been made. However, a much stronger factor was the improvement of global copper prices. This would not be the last time that Mongolia's economy and its resilience was put to the test. The Asian energy crisis of 1997 made the situation worse and subjected the country to economic stress again. The adopted changes of the early 1990s predetermined that Mongolia's economy undertook a new principal path for growth. External trade started increasing, especially with China, and inflation stabilized. Glimpses of future economic recovery began to appear on the horizon.

The mining industry, a stable platform for the country's economic recovery, was in dire need of investments, and these could come after the adoption of the mineral law. In collaboration with the World Bank (Mongolia joined the World

Bank membership in 1991), a group of experts was put together to work on the legislative documents. They created an eloquent set of mineral regulations designed to benefit all stakeholders, which includes the state, the owner of the mineral resource in the ground, investors, and potential developers of these mineral resources. Without this fundamental change, it was doubtful that any meaningful investments would come to the mining sector in the country.

1.4 LEGISLATURE AND INVESTMENT POSSIBILITIES IN MONGOLIA

The transition toward the market economy required radical changes to the country's legislature. The state had to adopt numerous legislative acts to define the rules of the "game," which were critically important for foreign investors. The Constitution of Mongolia served as a base for various judicial regulations. A series of legislative acts came into effect, for example, Environment Law, the Law on Foreign Investments, and the Mineral Law. Although the Great Khural adopted and enacted them at different times, the latter two laws turned out to be most critical in predetermining the course of the future development in Mongolia's mining industry.

Until 1989, all production capacities in the country, including mining, belonged exclusively to the state. State-owned companies such as Erdenet and MongolRosTsvetMet played a crucial role in the production of mineral resources in the country.

In 1991, a variety of legislative acts were adopted, including essential laws on enterprises, banking, bankruptcy, consumer protections, Labor Law, and so forth. Privatization started in October, which went through two stages. Initially, the government issued privatization vouchers, and then during the second step, it began trading these coupons for money. In 1992, the Great Khural of Mongolia adopted a new Constitution which established the right of private ownership in the country.

The work on amendments to the 1989 Law on Subsurface started in 1992. In March 1994, the government of Mongolia established the Ministry of Energy, Geology, and Mining by consolidating the ministries of Ministry of Fuel and Energy and Ministry of Geology and Mineral Resources. This created a single entity entirely responsible for the mineral resources of the country. The new Ministry consisted of the Geological Department, Department of Energy, Mining Department, and Department of Cooperation. The Geological Department was responsible for exploration.

In 1994, new amendments to the Mineral Law, taken from the Land Law, came into effect in 1995. These legislative documents reinforced state ownership of mineral resources in the country. Unfortunately, the new mineral-regulating act did not define long-term rights of mineral investors, leaving the area of long-term conditions to the government and the Ministry. Some historical factors influenced the way the changes were made, such as historical factors of the social past and conditions of the nomadic lifestyle in the country based on the joint social ownership of land. It was necessary to create a more confident legislative foundation for investors in the mineral sector, in which the state should guarantee the mining rights of the mineral deposit discoverers.

The Foreign Investment Law was initially adopted in 1990. It created a solid base for the state policy toward foreign investments in the country. From the moment of transformation toward a market economy, Mongolia diligently pursued foreign investments that had the potential to contribute to the economic stimulation and growth of the national economy. In 1993, new amendments to the Foreign Investment Law were enacted.

Indeed, in the future the law underwent additional changes. However, the basics and the principles have remained to this day. Some of the more important ones are guarantees to investors from nationalization and expropriations and to the stability of judicial base, allowing only changes for further improvement of the legislature and foreign investments in the country. Investors are free to select the economic type of activities and geographical areas of their interest, except for areas and zones where investment was prohibited. Mongolia guaranteed rights to all foreign investors irrespective of the origin of their country. Investors were given a stability guarantee for tax purposes depending on the size of their investment. Investments exceeding US$20 million dollars would be protected for 10–15 years.

The Mineral Law, as mentioned previously, changed several times since 1989. Changes of 1994 were promulgated in 1995. Most radical changes took place on July 1, 1997. This version of the Mineral Law was prepared with the assistance of the World Bank and incorporated the most advanced conditions in the area of mineral regulations from such leading mining countries such as Chile, Canada, and Australia. The law regulated all minerals in the ground except water, oil, and natural gas.

During the creation of the JV Erdenet–Magma and to the present, the state was the owner of mineral resources and defined the procedure of issuing licenses, their registration, and processes of resolving disputes. The state also determined the land for special use or protected areas, where the Mineral Law was limited or entirely lost its functionality and application. The primary regulator of the process was the Cadastre Office on geology and mining. The Agency on Mining and Geological Development was in charge of issuing licenses. No limitations on the participation of foreign investors in the development of mineral resources were envisioned.

Before applying for an exploration license, potential applicant could conduct reconnaissance, informing the Cadastre Office about their intention to perform such activity, indicating investors' representative officials, their names, and addresses with contact phone numbers.

The law envisaged issuing an exploration license by the Cadastre Office after 10 days from the moment of the application submission. Obtaining an exploration license was not limited to Mongolian citizens; anybody with credentials had access to the process. The permit allowed the owner to conduct exploration during the first 3 years, which had the potential to be extended for two additional 2-year periods. The size of the exploration area was set to be between 25 ha and 400,000 ha.

The licensed owner was obligated to prepare annual reports informing the Agency on Mining and Geological Development on exploration work performed and on plans for environmental protection. The environmental programs had to be submitted to governors of aimags and somon offices. It was not required to reduce exploration license ground as exploration license progressed its term. The license owner had the right to transfer the permit. The license fee was 5 cents per hectare for the first year, which doubled during the next 2-year period. During the fourth and fifth anniversary, the payment increased to US$ 1 dollar per hectare, and during the final 2-year term, the amount was US$ 1.5 dollar per hectare. The law did not require minimum work commitments. In case of success, the license owner had automatic rights to mine the newly discovered deposit. The license area had to be rectangular, with its sides in both meridional and latitudinal directions. The legislature established a mining term for 60 years, which could also be extended for another 40 years, constituting an overall period of mining granted for 100 years. The Mineral Law did not expressly limit the number of licenses that each license applicant could hold.

Naturally, the fee for mining license was higher. During the first 3 years, it was established to be US$ 5 dollars per hectare, US$ 7.5 dollars during the fourth and fifth years, and US$ 10 dollars per hectare for the final duration of the mining license. The relatively high amounts were established to curb potential speculations.

The owners of both exploration and mining licenses were obligated before the commencement of their work to receive approval from the relevant government authorities of the environmental protection plans for the activities they were planning within the licensee area. The owners of the licenses as indicated previously were free to transfer their licenses or use them as finance guarantees.

The financial conditions of 40% on profit and 20% on dividends were considered to be relatively high and marginally acceptable for business compared with the rest of the world. The royalty was set up at 2.5% level from the total production output. One of the attractive aspects was a free tax provision on all imports and exports. Besides, for the first 10-year period, all mining companies (except gold producers) received full tax relief for the first 5 years and then 50% discount for another 5-year period of the total tax relief stretch of time. Value-added tax on gold producers was at 10%, and it was 13% for other mineral producers. The state guaranteed to honor the fiscal conditions of the investment for a 10- to 15-year term.

Exploration and mining activities were required to be within the ecological guidelines defined by the Mongolian Environmental Law. A number of international laws supported the Mineral Law, for example, the Law on Protection of the Ozone Layer. The Ministry of Environment was given responsibility for the administration of environmental aspects of the Mineral Law.

To create funds for future recultivation and environmental restoration of ground disturbed by mining operations, the mining license holders were required to contribute a bond consisting of defined amounts of money.

1.5 JV ERDENET–MAGMA

At the beginning of the 1990s, many promising Mongolian cadres started arriving in the United States for training. The Colorado University in Boulder conducted a training program for many foreign nationalities, including the experts from Mongolia. A representative of Erdenet Tserennadmid, apart from attending courses, also made efforts helping the Erdenet management to establish contacts with copper producers in the Unites States, including Magma Copper.

Initially, the management of Magma Company was not attracted to the idea of investing in Mongolia. However, with time, the concept of cooperation with Erdenet gained some momentum. Vice President Bradford Mills played a pivotal role in rallying company support for the decision to take a closer look at the investment opportunities in Mongolia.

Initially, Magma's attention was directed at the Erdenet Enterprise. The management of Erdenet, namely the General Director Shagdariin Otgonbileg, had been attempting to attract investors who could help the enterprise implement several projects to improve production and processing of minerals, including low-grade oxidized copper ore stored in dumps.

In 1994, Magma Copper sent a team of experts to Mongolia with the purpose of reviewing the investment opportunities at Erdenet. They were offered various business opportunities from investment into a heap leaching project to brownfield exploration. The Magma representatives did not believe that the low copper grade dumps were sufficiently prospective for the company. However, they were attracted by the mineral potential for extension of the known mineral resources of the property.

After a detailed study of the mineral potential of the Erdenetiin district, experts recognized that the prospectivity for finding new large deposits with high-grade copper ore in this district were limited. However, based on the existence of hundreds of copper prospects known across the country, the experts concluded that the residual potential of the rest of Mongolia was still good. The team believed that the country with the relatively low level of historical exploration still had a reasonable potential for new discoveries. They recommended that Magma should consider a JV exploration program in Mongolia with a local partner well experienced in copper exploration and mining.

Erdenet apparently was the qualified candidate. A 50-50 JV was set up to equal share the funding and the risk. The underlying assumption was that these two companies could provide all the attributes required for a successful enterprise. Magma Copper was capable of contributing exploration personnel with a track record of copper discoveries and with experience in conducting modern, large-scale exploration programs. Magma also had the staff and expertise for running large-scale, world-class copper mines. Erdenet, on the other hand, had knowledge for conducting exploration in Mongolia and facilitating exploration work through its good networking relationships on both local and governmental levels.

Indeed, Erdenet-Magma JV was facing a daunting task. The venture had not only to find a relatively high-grade and sizable copper deposit but because of the remote setting and lack of developed infrastructure, it must also be a facility more extensive and more efficient than any owned at that time by either JV partner. Both partners realized how small the odds were of clearing these high hurdles.

JV negotiations advanced in a very tense environment with vigorous debates, several times risking to wander off the edge of a cliff—both partners were taking a new path in the creation of a JV. There were no previous analogs for anybody in these conversations. It was evident that to be successful, innovative approaches would have to be taken in finding mutually acceptable solutions. The process also required a great deal of trust among the partners. Brad Mills was leading the Magma delegation, whereas Shagdariin Otgonbileg was in charge of Erdenet. Finally, the partners reached a successful conclusion on April 1995 and officially signed the JV agreement in Ulaanbaatar.

The emblem of the Joint Venture Erdenet–Magma appeared in the form of a fast copper steed, rushing on the open green spaces of Mongolia toward new copper discoveries. This idea reflected the aspirations of both partners to achieve great success in the development of the copper industry of Mongolia.

In reality it was the first step towards the future discovery of Oyu Tolgoi. However, nobody could perceive and realize this at that time. Nevertheless, the determination to achieve success and make a significant difference was paramount among the participants.

Chapter 2

Initiation of Exploration—The First Regional Reconnaissance

Chapter Outline

Sergei Diakov *(Serguei Diatchkov, later Sergei Diakov)—Ph.D., former associate professor of the Peoples Friendship University, Moscow, Russia, an expert in structural geology of polymetallic, copper, gold, uranium, and nickel deposits. While with Energy Fuels from 1993 to 1995, he provided a foundation for the first Mongolian–Russian–American JV Gurvan Saikhan to explore and develop in situ leaching uranium deposits in Mongolia. Since April 1995, he became the Manager of the exploration program of Magma Copper in North Eastern Asia and General Director of the JV Erdenet–Magma. Consequently worked at BHP, Anglo Gold and currently with Anglo American as Group Discovery Advisor, United States.*

Discovery of Oyu Tolgoi. https://doi.org/10.1016/B978-0-12-816089-3.00002-0

Mark Sander—*Bachelor of Amherst College in Massachusetts in 1997, Master of Stanford University in California in 1981, and Ph.D. in 1988. Since 1988, he was a geologist of gold company Echo Bay, from 1990 to 1992; analyst of Magma Copper, since 1992; manager of the company on strategic planning; and director of strategic planning, since 1994. Mark was directly involved in the creation of JV Erdenet–Magma Copper. Since 1996, he was the Vice President of BHP on strategic planning. Currently, Mark Sander is the CEO and Director of Mandalay Resources working closely with Brad Mills.*

Eric Seedorff—*a graduate of the University of California, Davis in 1977, then Masters of Stanford University in 1981 and Ph.D. in 1987. From 1976 to 1986, Eric had short consulting assignments at mining companies Noranda, Anaconda, Climax Molybdenum, and Inspiration Mines. From 1987 to 1989, he worked in Chevron and from 1990 to 1991, in WestGold. In 1991, Eric started his career at Magma Copper initially as a geologist, then Chief Mine Geologist at Robinson, and consequently, Chief Geologist of Magma Copper. From 1996 to 1999, he was the Vice President of Mineral Resources in BHP Copper. Currently, Eric is a professor at the University of Arizona in Tucson.*

Dandinsurengiin Khishigsuren—*a graduate of the Irkutsk Polytechnic Institute in Russia in mining engineering. After graduation, he worked for the state company ErdeneMongol. In May 1995, Khishigsuren was appointed to the position of JV Erdenet–Magma Office Manager and Deputy General Director of the JV. From 2000 to 2002, he worked as the Head of Technical Team at the Cadastral office of MRAM. From 2004 to 2012, he worked at Vale as a Land Manager and Environmental Coordinator, and since 2015, Erdenet has employed Khishigsuren as the Head of Technical Department.*

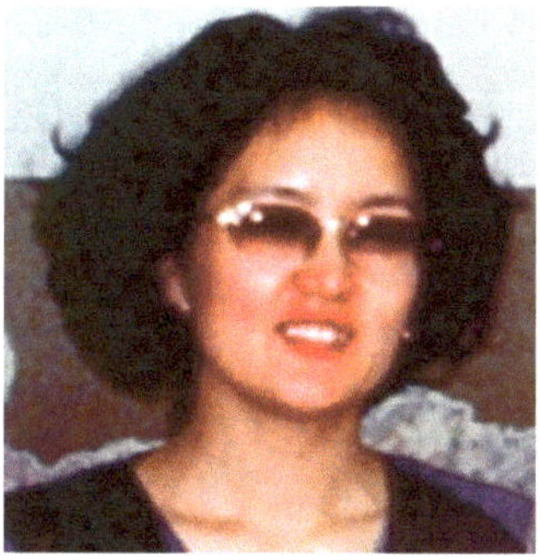

Gundegmaagiin Sarangerel *graduated from Mongolian State University; she was appointed to the position of JV Erdenet–Magma accountant in May 1995, initially on a temporary, and then in October 1995, on a permanent basis. From 2000 to 2004, Sarangerel worked as a senior accountant for Ivanhoe Mines. Currently, she works as a chief accountant for MineInfo in Mongolia.*

Dennis P. Cox—*famous American geologist, expert in models of mineral deposits, graduate of the Stanford University in 1956, worked for Anaconda in the ore district of Tintic in Utah State. From 1961 to 1995, he worked in the US Geological Survey; mapped copper deposits Tanama and Helecho in Puerto Rico, Ruby Creek, and Kennecott in Alaska; and studied structural control of mineralization at copper porphyry deposit Ajo in Arizona. Denis Cox together with Donald Singer was the editor of the widely known monograph "Models of Ore Deposits" (Cox and Singer, 1986), published by the USGS.*

Vitaliy I. Sotnikov—*doctor of science, professor of the Department of Mineral Deposits of the Novosibirsk State University, head of the research laboratory of ore systems of the Siberian Branch of the Russian Academy of Sciences, member of the International Academy of Mineral Deposits. Vitaliy Sotnikov is a laureate of the State Prize of the USSR, author of seven books and over 250 scientific articles, an expert in the field of ore formation and metallogeny, published numerous articles on copper prospectivity of Mongolia. He is now deceased.*

Gundsambuugiin Sanduijav—*a senior geologist of Erdenet, together with Czechoslovak geologists participated in geological exploration works at the Erdenetiin Ovoo deposit during the earliest stages. He participated in the assessment and calculation of mineral resource of Erdenetiin Ovoo deposits. He is a laureate of the State Prize of Mongolia for the discovery of Erdenet. A geologist with vast practical experience in prospecting and exploration for copper porphyry mineralization, as well as the operational intelligence on existing copper porphyry mines. He is currently retired.*

Dondogiin Garamjav—*a graduate of the National University of Mongolia, after graduating in 1967; he worked in the geological mapping expeditions. From 1968 to 1972, he participated in geological exploration works at Erdenet and had significant knowledge of copper mineralization in Mongolia as well as magmatic structures, in particular, the formation of alkaline complexes. In the mid-nineties, Garamjav was a member of the Mineralogical Museum of the Ministry of Geology. Since 1995, he was the consultant for the JV Erdenet–Magma, and later became a staff member of BHP as a senior geologist.*

Richard B. Leisure—*a consultant geologist. Richard provided consultancy to Magma Copper in gathering and analysis of geological data of copper mineralization in Mongolia, Tuva, and the Russian Far East. He actively participated in the preparation and execution of the first phase of the field reconnaissance activities of the Erdenet–Magma JV. Richard is currently retired.*

Galsan Jamsrandorj—*a geologist, graduate of the Irkutsk Polytechnic Institute in 1975, a candidate of geological and mineralogical sciences from the Moscow Geological Institute. From 1987 to 1994, Jamsrandorj worked in the Institute of Geology and Mineral Resources of the Ministry of Geology of Mongolia, and then as the Director of Geological Survey Department in the Ministry of Geology. Jamsrandorj is an expert in gold and uranium. From 1994 to 2000, he was the first Vice President and then the President of the private gold mining company developing Ovot Gold Deposit. In 2014, he became a member of the Mongolian Committee for Mineral Reserves.*

Robert P. Ilchik *has a bachelor's degree from the University of Nevada in Reno in 1976 and Master's degree from the University of California at Berkley in 1984, and then in 1990, Ph.D. from the University of California in Los Angeles. Bob is an expert in Carlin-style deposits of Nevada, copper porphyry, epithermal, VMS, and copper-gold systems of IOCG. During the time of JV, he was based in Tucson, but now Bob lives and works in Australia.*

Jeffry Lee—*Ph.D., and an expert in structural geology and tectonics. He has a bachelor's degree from the Pomona College in 1980 and Ph.D. from the Stanford University in 1990. Currently, Jeff is a professor of structure and tectonics in the Department of geological sciences at the Washington State University in Ellensburg. He published numerous articles and papers on structure and tectonics of continental orogenic belts and extension systems.*

Tseekhuugiin Tsend-Ayush—*a geologist, graduate of the Irkutsk Polytechnic Institute of the 1975 year. He started his career in Tsagaan Suvarga Geological Expedition. From 1985 to 1987, he worked as the Chief Geologist of the Khovd Geological Expedition. He has expertise in gold, copper, and silver. Tsend-Ayush is an experienced geologist with significant practical experience in the field; he is currently a geologist-consultant of the company "MMNS."*

Tumur-Ochiriin Munkhbat—*a geochemist, graduate of the L'viv State University, Ukraine in 1991. He worked in the Institute of Physics and Technology in Mongolia. From 1998, he is a permanent employee of BHP Minerals Mongolia. Currently, Munkhbat is the Deputy Director of the Department of resources and strategy innovation of Oyu Tolgoi Company.*

Gonchigiin Oyun—*geochemist, graduate of the L'viv State University, Ukraine in 1992. From 1997 onward, she worked at BHP Minerals in Mongolia, and then she worked at the company QGX of Mongolia. In 2012, Oyun founded the company HHP Mongolia.*

John Barry Prescott—*a graduate of the University of New South Wales with Bachelor of Commerce degree. John started his career at BHP in 1958, initially as a trainee in the steel unit at Newcastle. Subsequently, he rose to the level of an Executive Director and the Chief Executive of BHP. In this post, he remained for 8 years from 1991 to 1998. Under his leadership, some significant events occurred in BHP, including titanium sands project Beenup in Western Australia, the pellet production plant in Venezuela and a similar plant in Western Australia, nonconfirmation of reserves of oil fields in Vietnam, and finally, the acquisition of Magma Copper. All these resulted in US$ 10 billion loss. On March 4, 1998, the Board of Directors forced John to resign from his post.*

Jeremy Ellis *was born in England, received his education in Western Australia. He managed production units in BHP: he was the Manager of the Planning and Development Division, and later the Manager of the Australian steel units of BHP. From 1989, he served as the Executive Director of BHP unit Utah International, and since 1991, he was the BHP Director and head of BHP Minerals. In 1997, he was appointed as the Chairman of the Board of Directors of BHP, replacing his predecessor, Brian Lotona, the former boss of John Prescott. He held this post until the year 1999, when Don Argus succeeded him in this position.*

Dick Carter *was born in Newcastle, Australia. He graduated from the University of New South Wales in 1964 and Harvard Business School in 1984. Dick began his career in 1960 as a BHP probationer at the steel plant in Newcastle. He held various positions in iron ore to the managing arm of BHP and from 1996 onward took the reins of BHP Minerals from Jerry Ellis, becoming the Executive Director of BHP Minerals. In 1997, Dick ended his 37-year career in BHP by taking responsibility for failures in the iron ore unit, in particular, for the fiasco of building factories of iron pellets.*

Hugo Dummett *was one of the primary motivators for diamond discoveries in Canada. Hugo possessed extensive experience and expertise in geological prospecting in the copper porphyry field. He was born in South Africa. In 1965, he received a B.A. degree from the University of Witwatersrand and then a Master's from the University of Queensland, Australia. In 1977, as a senior geologist, he worked for Superior Oil, where he began exploration for diamonds. After the discovery of the Lac de Gras by Dia Met Company Hugo persuaded BHP to support the project. Hugo served as the Vice President of BHP Minerals. In appreciation for his achievements in exploration, Hugo received high awards. After BHP, Hugo worked for Ivanhoe Mines. In 2002, the year of his tragic death, he was the President of the Society of Economic Geologists.*

Donald J. Schissel *received his bachelor's degree from College in Massachusetts, Amherst. Don graduated with Master's diploma from the University of Montana in 1981. From 1974 to 1979, he worked for Stillwater Mining Company and participated in the exploration of one of the largest deposits of platinum group elements in North America. From 1981, he was with BHP. At the time of the events, Don held the position of Regional Manager of Northern Eurasia BHP based in London. The Central Asian program was part of the North-Eurasian region.*

2.1 SETTING UP THE JOINT VENTURE

One day in February 1995, senior geologist Sergei Diakov's desk phone rang in the office of the American company Energy Fuels in Denver, Colorado. One of the local recruiters was calling Sergei with an unusual request to assist in identifying a profoundly experienced Russian-speaking geologist with experience of exploration work in Mongolia.

Energy Fuels was involved in uranium exploration in Mongolia under the Mongolian–Russian–American Joint Venture (JV) named Gurvan Saikhan. The venture was created in 1994 to study in situ leaching method of uranium extraction in the Mongolian Gobi. In addition to uranium ventures in Mongolia, Energy Fuels also considered gold opportunities. Several gold prospects attracted the company's attention. Sergei Diakov together with Gorol Dimo conducted evaluation studies.

This was when communication with Magma Copper started for Sergei. Magma was looking for an experienced geologist for conducting exploration in Mongolia. Initial connection was through phone with Mark Sander and Eric Seedorff. An invitation to lead the exploration program in Mongolia was very attractive. The conversation that started in February consequently leads to an invitation to come to Arizona for an interview with Magma Copper in Tucson. After a meeting with Brad Mills, Executive Vice President of the Company, Sergei was offered a position as leader of Magma's exploration program in Central Asia, including Mongolia. He was told that it was necessary to start operating by April 1. The position was in Tucson, Arizona to which Sergei had to relocate from Denver.

Richard Leisure, local senior geologist in Tucson, contracted by Magma to collect data on Central Asia, was gathering the information from numerous public sources on copper prospectivity of Russian Tuva, Far East Russia, and Mongolia as well. Although the data collected still required analysis and prioritization, initial work was already underway, and Richard put the first bricks in the foundation of the Mongolian exploration program.

According to the JV Agreement terms signed by Magma and Erdenet, the Management Committee that included three representatives from each side directed the JV. The Committee was empowered to review and approve the working programs and the budgets for their execution. The General Director of the JV Erdenet—Magma was responsible for the approved programs' implementation. The Management Committee of the JV Erdenet—Magma included Sh. Otgonbileg (later replaced by Ts. Sanjii), and Munkhbaatar and M. Erken. Brad Mills, Mark Sander, and Eric Seedorff represented Magma Copper in the Management Committee. The first meeting took place in Ulaanbaatar where the Management Committee appointed Sergei Diakov to the position of General Director.

The JV General Director was in charge of the day-to-day operation of the JV, maintaining regular bookkeeping, providing relevant reports about its financial and other types of activities that were required by the legislature in Mongolia. For this, it was necessary to register the JV, rent an office, hire office staff, start bookkeeping, etc., all according to the legal requirements for business enterprises operating in the country.

Immediately, after signing the JV Agreement, Magma and Erdenet started the registration process and preparations for the field season. In May 1995, the LLC «Erdenet—Magma» was formally registered in Mongolia under the registration number 2072548. Besides setting up the JV, it was also necessary to start collecting the geological data and initiate preparations for the fieldwork.

Dandinsurengiin Khishigsuren, a mining engineer by profession, was appointed to the position of the Office Manager and Gundegmaagiin Sarangerel was hired to the position of the JV accountant, initially temporarily, and then on a permanent basis. Both of these employees approached their new job very responsibly and within a short time managed to organize the work of the JV Office so that the Erdenet—Magma JV could begin functioning efficiently.

Sharing of accumulated technical knowledge and experiences with each other allowed the JV partners to run the JV more efficiently and find better technical solutions for their operations. The technical discussions and visits to the experimental sites accompanied the Management Committee's meetings. During one of such meetings in Tucson, the delegations of Erdenet asked for an opportunity to visit Magma production facilities, including several experimental sites for in situ leaching operation to produce copper cathode (Fig. 2.1).

The management of Erdenet was interested in studying Magma's experience of copper ore processing. In addition to concentrate production, Magma was also involved in a heap and in-situ leaching of copper oxides to produce

FIGURE 2.1 San Manuel mine, a polygon of underground copper leaching process. Left M. Erken, Chief Geologist, S. Otgonbileg, CEO of Erdenet, the rightmost Sergei Diakov.

copper cathode and rod. Dumps of low-grade oxide copper ore at Erdenet were seen as a source of compensation for declining at depth copper content in the ore of the Erdenetiin Ovoo deposit. The increased hardness and abrasiveness of the ore material, which consequently required more massive energy consumption, cause increased wear and tear of the equipment, eventually affecting the cost price of production of the copper concentrate. Looking for new solutions was critically necessary. One of these solutions involved copper production using leaching technologies for copper extraction from low-grade oxidized ores, and later possibly molybdenum production from the low-grade dumps. Heap leaching with copper extraction through electro winning to obtain copper cathode was in the discussion.

During the JV Management Committee meeting in Tucson, Arizona, Magma offered to let their partners from Erdenet to visit a pilot test site for extracting copper using the in-situ leaching method. Magma had accumulated extensive experience in methodology for processing of oxidized copper ores. It actively used heap and in-situ leaching methods. Moreover, experiments were also conducted for heap leaching from primary sulfide ores, such as at Pinto Valley near Globe–Miami in Arizona. Of particular interest was a testing ground for the extraction of copper from the oxidized ore in the slopes of the open pit San Manuel, which was the site where the tour for the Mongolian Management Committee members was arranged.

It was interesting that despite low porosity of the host rocks at San Manuel, Magma experts developed an efficient technology for in-situ leach of copper recovery from rocks in the ground without mining. Discussion of the original technological solutions, accumulated by both partners through their experience of copper production, created a solid base for fruitful mutually beneficial cooperation between partners in the JV Erdenet–Magma.

2.2 DATABASE RESEARCH AND SELECTION OF PROSPECTS FOR FIELD EVALUATION

In 1995, some 600 copper mineral occurrences of varying genetic types with copper manifestation on the surface were known in Mongolia. Only two of them were copper porphyry deposits, one of which Erdenetiin Ovoo of Permian–Early Triassic age, was in production, and the other Tsagaan Suvarga, of Devonian–Carboniferous age, was still in the exploration stage. In addition to these two copper deposits, there were numerous vein-type mineral prospects of hydrothermal, sediment-hosted, volcanic massive sulfide origin, as well as copper skarns mapped and recorded by previous state exploration. All information for these occurrences was stored in the geological archives controlled by the state. Mongolia's leadership recognized that open access to geological information helps to attract foreign investment in the mining sector. This enabled potential investors to examine geological data and to choose the prospects that would be most attractive for their investments. The Magma–Erdenet JV team also received free access to the geological archives, including reports, maps, geophysical and geochemical data.

The team had a task to complete the data review and come up with recommendations for field visits in a very brief period, as long as 1 month at maximum. The mission was to review previous geological reports and select a range of known copper occurrences, which could still have an attractive geological setting for copper deposits and had some upside exploration potential for further expansion and quality improvement as a result of additional exploration. At that time, we did not limit our attention to certain types of copper mineralization. Intuitively, we thought that porphyry and sediment-hosted style of mineralization would warrant the most significant in size, whereas VMS and copper skarns would provide the richest regarding their potential copper grades. Nevertheless, at the initial stage, the decision was made to select a reasonably wide range of copper prospects for our field inspection. After collecting field observations, we were planning to come back to the question on which type of copper deposits the team should be focusing to deliver the JV expectations and could have a potential to become a significant size copper mine.

The plan was that after field evaluation of the identified prospects of interest with the geological potential for discoveries, we would rank them and prioritize based on their potential to yield a significant copper discovery. However, the ranking process also had to take into account some critical factors, most notably the likelihood of obtaining a license and securing the ground as well as the project's proximity to the existing infrastructure.

2.3 ASSEMBLING RECONNAISSANCE TEAMS AND PLANNING FIELD CAMPAIGN

A technical group consisting of G. Sanduijav, D. Garamjav, G. Jamsrandorj, and Ts. Tsend-Ayush was appointed to study the data in the Geological Archive.

The principal technical representative of Erdenet was senior geologist G. Sanduijav. The General Director of the JV handled all aspects of selecting and hiring expert geologists, including contract negotiations. Care was taken to assemble a well-rounded, diverse, international team of exploration geologists with extensive experience in copper exploration and a broad knowledge of copper deposit styles including the conditions under which they were formed. The local geologists brought with them the added expertise of copper exploration in the Mongolian terrain.

The geologists were divided into two teams. Team 1 included Richard Leisure, Dennis Cox, Vitaliy I. Sotnikov, Gundsambuugiin Sanduijav, and Dondogiin Garamjav; Team 2— Sergei Diakov, Galsan Jamsrandorj, Robert Ilchik, Jeffrey Lee, and Ts. Tsend-Ayush. Tumur-Ochiriin Munkhbat and his wife Gonchigiin Oyun joined team 1 later on.

The planned exploration program was very ambitious for the short 3 months long field season lasting from the beginning of June until the end of August. A total of 75 prospects had been selected and scheduled for site visits, which were spread out over an extensive area almost 1,200 km east-west and more than 600 km north-south. We estimated that on average, each prospect might require two field days to evaluate, plus an additional day to relocate to the next copper occurrence.

The JV office began active preparations for the startup of the field activities. In preparation for the field campaign, the team required purchasing of various field equipment, field supplies, field gear, mobile kitchen equipment, and different types of safety gear. Office Manager Khishigsuren shouldered many of these responsibilities while Sarah and two cooks procured food supplies and various cooking amenities. This enabled the geologists to focus on the technical aspects of the program.

The General Director set the goal for the field teams to depart Ulaanbaatar in early June allowing sufficient time to visit at least half of the scheduled prospects before the Naadam holiday. Three weeks were allocated for a midfield season break. The second half of the field program was expected to begin on July 28, shortly after Naadam.

The first team surveyed the northern and central part of Mongolia from Erdenet to Darkhan (Orkhon-Selenga zone), where nine prospects were located. The second team reviewed the west and central parts of Mongolia from Bayankhongor to Gobi-Altai (Bayankhongor zone), which contained 15 of the prospects. During the second half of the season, the first team was expected to check the 25 copper occurrences in the south of the Gobi Desert between Sainshand and Dalanzadgad. The second team was to review 26 prospects in Western Mongolia between the Khovd and Gobi-Altai region.

It was essential to verify the findings reported by the Mongolian, Soviet Union (mainly Russian), and Eastern European countries' geologists. Some members of the 1995 teams had the opportunity to directly work or at least visit a number of sites in the past. For each prospect, it was essential to confirm the presence and extent of previously identified copper mineralization, select necessary geochemical samples, and to attempt to assess the possibility of discovering additional mineralization, which could have the potential for economic copper possibly with credits of gold, molybdenum, and other associated nonferrous metals.

Various logistical difficulties arose, which occasionally hampered the progress. One of the challenges was the difficulty of communicating in Mongolian, English, and Russian languages. We recall one case when a local shepherd in western Mongolia was only adequate in the Kazakh and Russian languages. Typically, a conversation with the Mongolians went through translation from English or Russian into Mongolian; however, in this case, the communication took place in the opposite direction.

The topography over most of the project area is flat to rolling. The exceptions are the Altai Mountains in the west, with peaks of up to more than 4,000 m above sea level, the slightly elevated Khangai Mountains in the central part, and the steep Khentei mountain range along the border with Russia. Mongolia has an extensive network of unpaved—mostly dirt roads which, over this relatively benign terrain, make it possible to drive off the road for considerable distances allowing relatively easy access with four-wheeler drive vehicles. Asphalt roads are primarily confined around Ulaanbaatar and a few relatively sizable local towns such as Darkhan, Erdenet, etc.

Weather conditions, strong winds, and torrential rains, in particular, provided safety challenges and occasionally obstructed progress. Three tents were destroyed in the first phase. Flash floods can occur during the summer months. The team leaders had the responsibility to select campsites that were sheltered as much as possible from these elements.

2.4 REGIONAL GEOLOGY OF THE RECONNAISSANCE AREA

Mongolia is built of a complex of island arc accretion prisms, which formed as a result of the subduction of oceanic crust beneath the Central Siberian continent. The oldest of them developed in the Late Proterozoic period (Vendian) and Early Cambrian period and included fragments of ancient

Precambrian of the Siberian platform. They consisted of deep oceanic sediments and basaltic volcanic rocks primarily developed in northwestern Mongolia. Other prisms formed in the early and middle Paleozoic period and then docked to the Siberian platform from the South. These blocks are likely transient conditions of the oceanic crust to the marginal continental regime. Along southern Mongolia, these rocks were intruded by numerous granitoid intrusions of Devonian age. In the Carboniferous and Permian, the volcanic arcs of the Andean type were developed with several back-arc depressions between the blocks of more ancient rocks. Intrusions of Carboniferous–Permian age subsequently intruded into the volcanic sediments and subaerial molasses accumulated in these sedimentary basins.

The island arc magmatism ended in the Late Triassic and Early Jurassic periods with the formation of subaerial lavas and tuffs, as well as intrusions of the alkaline composition. In the late Jurassic period, sedimentary basins originated, which were filled by conglomerates and coarsely fragmented sediments. Broad marsh ponds, in which shallow marine sediments accumulated, formed in Cretaceous. At this time the final landscape of modern Mongolia developed. Cenozoic sedimentary platform cover formed in a series of shallow lacustrine basins across the country.

During the evolution of the geological development of Mongolia, several types of copper deposits were formed (Berzina and Sotnikov, 1991). They were volcanogenic-sedimentary from Early Cambrian to Middle Paleozoic; copper–nickel mineralization associated with Middle Paleozoic intrusions of basic composition; strata-bound copper deposits of sedimentary origin formed in continental molasses and red bed sediments of Late Paleozoic Jurassic age (authors note: examples of descriptions of such deposits had not been found). The list included porphyry copper deposits and skarn deposits; vein-type deposits, as well as metasomatic deposits within transitional zones from the island arcs to the continent in favorable conditions of formation of continental volcanic arcs of Andean-type in Carboniferous and Triassic.

2.5 EXECUTION OF RECONNAISSANCE FIELDWORK

The Mongolian team started their preparation in Erdenet in late April, and the American team in Tucson began their preparation in May. In early June, both field crews were dispatched to their relevant designated areas of field reconnaissance. Significant effort and time went for development of the dynamics in the execution of the reconnaissance trips by both teams. There were numerous instances of logistical and technical difficulties experienced by the groups. At the end of July, after the Naadam holiday, the teams gathered in Ulaanbaatar to resume the reconnaissance program. The first field team went to the south, and the second field team traveled to Mongolian Far West to Khovd. The all-terrain truck GAZ-66, headed by Ts. Tsend-Ayush, went to Western Mongolia in advance. The rest of the team traveled by plane from Ulaanbaatar to Khovd.

According to the General Director's plan, both teams would complete traversing and return to Ulaanbaatar in early September.

Eric Seedorff attended the final portion of the fieldwork in the south. On September 8, the two teams gathered in Ulaanbaatar to review the results of their field observations and discuss the preliminary results of the reconnaissance work. The heads of both field teams made their presentations, and then there was a discussion of the reviewed information to determine the follow-up work program.

Of course, the final processing of the data and analysis of the samples collected in the field required some additional time. The JV management well understood that the conclusions of the workshop on September 8 were preliminary and deserved further confirmation based on the analytical results from the collected samples. All geochemical samples taken by both teams were prepared and processed in Ulaanbaatar and then sent to Vancouver for analysis at the Chemex Laboratory, but all this needed some time for completion.

The JV Erdenet-Magma Management Committee planned its next meeting in Tucson for the 18th to 20th of September. Although it was not expected that all test results would arrive in time for this meeting, nevertheless, the General Director planned to discuss the preliminary findings of the reconnaissance work with the members of the Management Committee to map out plans for the possible continuation of the work in the next stage. Upon receipt of the results of analyses, conclusions and recommendations should have been refined, but at this stage, in the hands of experts, there was enough material to make preliminary findings of how and where to continue the work at the next step and where the JV would need to focus its attention in the future. The list of issues included funding for the next year, which the Management Committee had to consider, review, and discuss, and then making a decision for approval.

The next Management Committee meeting was scheduled for the end of November. By this time, the results would undoubtedly arrive from the laboratory. Therefore, the leadership of the JV could approve the proposed work plan and budget for the next year, in the September meeting. In the meantime, both JV field teams (see the core of Team 1 on Fig. 2.2) were working on data preparation and finalizing the field reports.

2.6 DISCUSSION OF RECONNAISSANCE PROGRAM RESULTS

During the field season of 1995, a total number of 73 prospects from the selected 75 prospects were field checked. The characteristics of the prospects are summarized in two tables in the appendices at the end of the book and also on Fig. 2.3.

FIGURE 2.2 Participants of the JV Erdenet–Magma reconnaissance team. In front (from left to right): Sanduijav, V. Sotnikov, R. Leisure, D. Garamjav, and D. Cox in the back, photo 1995.

FIGURE 2.3 Map of copper prospects visited by Erdenet–Magma field teams during initial reconnaissance campaign. (1) Copper prospects, (2) Administrative centers, (3) Railroad tracks, (4) Lakes.

Based on the first team's field observations, Shuteen and Ih-Shanh were recommended for further study. Shuteen was located in the South Gobi, approximately 65 km from the village of Mandakh and about 60 km northwest of the Tsagaan Suvarga deposit. Shuteen consists of intensely argillized volcanic rocks of Carboniferous age that extend for a strike length of 12 km with a width of 4–6 km. Monzonite intrusions of Carboniferous age and alkaline granites of Permian age intruded these volcanics. In the area of alteration, four copper occurrences were found. Shuteen was repeatedly investigated based on the reports from Chovan (1983), Korim (1984),

Shabalovskiy and Garamjav (1984), and Delgertsogt and Daramsenge (1992). The state had mapped the prospect at scales of 1:25,000 and 1:10,000. Underlying the volcanics is a package of Lower Carboniferous agglomerates, intermediate lavas and tuffs, and sandstone, occasionally with conglomerates. The field teams verified the geological situation at the prospect sites they visited (Fig. 2.4).

Zones of alunitization and pyrophylitization occur within the sericitized volcanic rocks. These zones are visible on the Landsat images. Many discontinuous ring structures emphasized the volcanic nature of the prospect and highlighted by the alteration zones. Previous trenching revealed copper sulfide mineralization of up to 0.5% copper grade. The Har Tolgoi prospect had been drill tested. The exploration drilling intersected low-grade copper mineralization of 0.2% (Chovan et al., 1983).

Geochemistry indicated anomalous copper, molybdenum, zinc, and silver within the hydrothermally altered rocks. Previously conducted IP geophysics identified several anomalous zones with low-intensity IP and low chargeability of 4%–6%.

The team concluded that these results were positive. The presence of significant zones of sericitic and argillic alteration suggested the potential for discovery of copper-porphyry deposits possibly with zones of secondary supergene sulfide enrichment. The presence of zones with advanced argillic alteration was also considered encouraging for the possibility of enargite veining, which would further enrich the copper grades.

It was proposed to apply for a license area of 2,700 square kilometers, covering a large part of the volcanic belt including the Mandakh mineralization in the northern part of the license area. If successful, it was suggested to conduct a detailed review of the previous work and study remote sensing

FIGURE 2.4 Experts of the first field team on the outcrop of Mandakh, photo summer 1995.

imagery (Landsat and SPOT). If positive, this would be covered with selective detailed mapping of the most attractive sites followed by geophysical and geochemical surveys.

Ih Shankh was also located in South Gobi within soum Tsogtsetsii at a distance of 30 km from the nearest village of Barunsu, the administrative center of the soum. In 1969, a magnetic survey had been followed by mapping at scales of 1:200,000 (Адъяа, 1971) and 1:50,000 Худэрбат (1981). In addition, Дашцэрэн (1972) and Goldenberg (1978) conducted their research in the area.

The district is composed of Lower and Middle Carboniferous age volcanogenic-sedimentary rocks, which are overlain by a sequence of Late Carboniferous to Middle Permian molasse continental sediments and volcanics. These rocks were intruded by intrusions of monzonites and syenites as well as by the granites and granodiorites of Permian age.

Within the area, several zones of advanced argillic were identified. This alteration extended for about 20 km from west to east in the corridor of about 10 km wide. Within the altered zone, intensely silicified plugs stood out in relief about 50 m above the argillic rocks. Typically, such topographic "pedestals" had slightly elevated gold and silver content up to 2.5 g/t, as well as arsenic and molybdenum. It was recommended that an application for the 3,300 square kilometers area be submitted for detailed prospecting and exploration.

The most exciting prospects visited by the second team were Shirt and Airag Uul. The Shirt Prospect is located in Bayankhongor aimag. Previously, in 1982, a group named Tekhnoexport had geologically mapped this region at a 1:200,000 scale. According to the map, Lower and Middle Devonian rocks dominate the area. The lower Devonian sequences are composed of basalts, rhyolites, and limestones, while the Middle Devonian is composed of conglomerates and sandstones. The maps indicate an unconformity between these two sequences. These packages of rocks are intruded by Devonian granodiorites and diorites, which contain some copper–lead–zinc mineralization.

The Lower Devonian meta-rhyolites were found to be intensely deformed resulting in the development of schistosity and isoclinal folding. Along the contact with the limestones, which are also folded, and altered to marbles, there was a zone of mineralization with malachite, azurite, and chrysocolla. At the Shirt site, there were numerous small trenches and excavations up to 10 m in diameter, which were indicative of historic copper mining from an iron cap. Samples taken from the iron cap contained high levels of copper ranging from 1.3% to 5.6% with gold credits ranging up to 3 g/tonne and minor zinc from 0.09% to 0.3%, less than 0.1% lead, and 0.004% molybdenum. Remnants of drill core from the previous exploration demonstrated that this mineralization extended to depth.

The JV experts interpreted that the mineralization was volcanogenic-sedimentary in origin with well-developed iron caps. Only limited exploration was conducted on this prospect in the past and indications are that the Shirt mineralized zone could include several sulfide bodies carrying gold mineralization associated with iron caps, which would serve as a good "sweetener" supplementing the copper–polymetallic mineralization at depth.

The JV team recommended that the JV should obtain exploration licenses over the prospective geology of Shirt. This would be followed with detailed reconnaissance over the licensed ground and selective mapping at the scale of 1:10,000 over the areas of interest. Over the mineralized and altered parts of the structure, or where the structure is unexposed, the JV would conduct a complex of geophysical and geochemical surveys to determine reliably the nature and the extent of the copper–gold mineralization. The Erdenet–Magma reconnaissance teams implemented this exploration strategy over all the exploration licenses.

The area of Airag Uul is located in Zavkhan aimag 60 km west of Ulyastai. The prospective area of interest is 68 km long and 22 km wide and included several mineral occurrences such as Zos Uul, Airag Uul, Zambain Khudak, etc. Geologists V. Samozvantsev, V. Golyakov, and others conducted geological mapping at a scale of 1:200,000 in 1978–81. At the time of the second team's visit, the Khubsugul Geological Expedition conducted a geological mapping program over the area at a scale of 1:50,000. Reconnaissance team 2 visited the area (Fig. 2.5).

The area fell within the Nuuriin structural-formation zone of the western part of the Central Mongolian Massif. The Massif consisted of three complexes: Upper Riphean geosynclinal rocks (serpentinites, lherzolites, pyroxenites, and hornblendites), Upper Riphean orogenic rocks (slates, plagioamphibolites, gneisses, phyllites, quartzites, and marbles), and the Upper Riphean-Lower Cambrian subplatform varieties of rocks (metavolcanics and

FIGURE 2.5 Field camp of the second reconnaissance team.

FIGURE 2.6 Telephone communication was maintained through Inmarsat satellite connection. Jeffry Lee is listening; photo summer 1995.

metasediments). Medium and Late Paleozoic intrusions of gabbro, tonalites, quartz diorites, and granites intruded all these complexes. B. Samozvantsev believed that the copper mineralization here formed in association with Riphean greenstone volcanic belts. While in the field the teams maintained connection through satellite Inmarsat phones (Fig. 2.6).

Historically, more than 40 points of copper mineralization had been identified in the area. Seven of these were inspected during the field visit. With the compliments of the Khubsugul Geological Expedition, the Erdenet–Magma JV team was able to examine a number of the Expedition's trenches dug on some of the prospects. At Airag Uul, copper mineralization grading from 0.22% to 1.5% was observed in the zones of silicification containing veinlets of quartz–pyrite in limonitized shales. Copper was also present in the Airag Uul zones of pyritization. At the Zambain Khudak, mineralization is associated with skarns developing in the limestones and calcareous sandstones. The skarns, composed of an assemblage of garnet–epidote–biotite–magnetite, contain disseminations of chalcopyrite and patches of malachite at their margins.

Due to limited bedrock exposure and time constraints in the reconnaissance area, the Erdenet–Magma JV team concluded that establishing an exact nature of the mineralization based on surface observations was too uncertain at that moment. The team made a preliminary conclusion that copper mineralization was most likely related to skarn development and feeder zones of volcanogenic-sedimentary lenses within metavolcanic rocks.

Based on geological potential and proximity to infrastructure, Shuteen and Ih Shankh comparatively looked more attractive than Shirt and Airag Uul. Both Shirt and Airag Uul were interpreted as volcanogenic-sedimentary deposits of the Kuroko type. Based on the US Geological Survey database by D. Singer and D. Cox (Cox and Singer, 1987), these types of deposits on average gave ore bodies of 17,000 tonnes of copper (or similar 240,000 tonnes of copper for 10% of the largest deposits of the entire family of base metal massive sulfide deposits). While for copper porphyry deposits, such ratio was determined by 600,000 tonnes of copper averaged and 5.4 million tonnes for 10% of the largest known deposits of the porphyry type. From this, it appeared that clear preference should be given in favor of the porphyry copper prospects.

Undoubtedly, accurately determining the probability of finding deposits at the above-indicated prospects was difficult due to multiple critical factors and limited information available for our judgment at that moment.

Based on the information collected from the reviewed prospects, Shuteen and Ih Shankh appeared to be associated with extensive zones of argillic and sericitic alteration. In Mongolia at that time, only the Erdenetiin Ovoo and Tsagaan Suvarga deposits were known to have significant zones of argillic/sericitic alteration. The remaining prospects interpreted to be porphyry style were either

- situated at deeper erosional levels where only potassic zone was preserved, or even at deeper levels where the phyllic-argillic alteration shells were eroded away;
- or these zones of alteration did not develop large-size volcanic centers;
- or the deposit style was just misinterpreted.

The interpretation of Shirt and Airag Uul being volcanogenic-sedimentary deposits of Kuroko gave rise to numerous questions. From the 425 mineral deposits of this type in the USGS database, only four had carbonate strata overlying a sequence of mineralized volcanic rocks. It was entirely conceivable that the genetic model for these kinds of deposits was misinterpreted.

Ultimately, the reconnaissance team decided in favor of the license application for the Shuteen, Ih Shankh, Shirt, and Airag Uul prospects.

2.7 REVISION OF EXPLORATION MODEL

By mid-September, all prospects scheduled for field evaluation had been inspected and sampled. Both teams experienced challenges either with accessing roads or with the adequate location of the prospects on the ground. A bigger problem was the lack of exposure in some areas, which resulted in spending extra field time searching for outcrops. Owing to inadequate exposure, it was not possible to determine the genetic origin for a number of prospects with any degree of confidence.

Following the return of the teams to Ulaanbaatar, the General Director called a technical meeting for the reconnaissance geologists to present the results of the fieldwork and to recommend a follow-up program.

The JV team fulfilled the task of examining most of the selected prospects during the 1995 field season. It was deemed that the support staff consisting of drivers, cooks, and field assistants made a significant contribution toward the success of this phase of the program and their strict compliance with the company's safety requirements ensured that there were no significant safety incidents.

During the presentations, there was a great deal of discussion around the question which types of copper deposits would likely deliver the most attractive copper discoveries. For example, vein-type copper deposits could not meet the ore tonnage thresholds required for an economic deposit in this environment. Porphyry types of deposits certainly had a necessary magnitude of size, but only if copper grades would exceed 1%. This could only be achieved if the deposit contained a good zone of secondary sulfide enrichment or a skarn zone or elevated values of by-product metals such as gold and molybdenum. Porphyry prospects in Northern Mongolia in the ore district of Erdenet contained some enrichment, although at a minor scale. Therefore, it was concluded that the identification of a copper-porphyry target with secondary enrichment in Northern Mongolia was very unlikely.

Dr. Dennis Cox, given his extensive experience in Arizona, proposed to focus on the model of copper-porphyry deposits with secondary sulfide enrichment or chalcocite blankets in which the copper content depending on the number of cycles of the formation zone of enrichment could increase two to three times compared to the copper content in the primary ores. According to Dr. Cox, the exploration criteria consisted of the presence of porphyry intrusions associated with large zones of quartz-argillic alteration zones, the presence of leached stockworks, and vugs or voids with remnants of copper, at the surface associated with limonites and hematite. This manifestation of leaching points to the process of weathering resulting in the transfer of copper from the surface to lower horizons. With this model, the presence of copper sulfide (chalcopyrite) at the surface is a negative feature. Not all participants immediately agreed with this concept. Nevertheless, Dr. Cox's conviction eventually prevailed. This idea was not entirely new. However, in the previous search for copper in Mongolia, it had not been thoroughly tested.

Based on this new concept, the experts recommended the following prospects (in order of their preference): (1) Shuteen, (2) Ih Shankh, (3) Shirt, and (4) Airag Uul. Zost Uul, Bayan Bulag, and Khuh Adar followed these in the list of priorities.

2.8 SUBMITTAL OF EXPLORATION APPLICATIONS

Applications for exploration licenses were filed in the Mongolian Geological Survey on November 15, 1995. Geological Survey rejected the applications for Ih Shankh, Shuteen, Shirt, and Airag Uul due to ongoing government program with JICA—Japanese Agency for mineral resource studies. In their place, three other prospects—Zost Uul, Bayan Bulag, and Khuh Adar—were submitted and accepted by the Geological Survey of Mongolia.

The events that began to develop at the end of the year 1995 were destined to affect the further progress of the JV radically.

2.9 ACQUISITION OF MAGMA COPPER BY BHP

News about the acquisition of the American mining company Magma Copper by the Australian multidisciplinary mining company BHP suddenly began to spread around the world in December 1995. This event would radically disrupt the Erdenet-Magma JV.

The purchase price of Magma was US$ 2.4 billion. This acquisition promised BHP would become the second largest copper producer in the world after Chile's Codelco. With Magma shares being US$ 21.325 dollars, BHP offered an excellent premium with a total price of US$ 28 dollars per share, i.e., 31% increase over the price of the shares of Magma, making it very attractive to the shareholders of Magma Copper. The consent of the shareholders was not required for BHP; the principal shareholder of Magma Warburg Pincus Capital Company owned 26% of the company, which was entirely in agreement with the terms of the deal. In addition to strengthening its production capacity for copper, this acquisition also enabled BHP to become a fully integrated copper company. The flagship of the copper production of BHP was the Escondida mine, which produced copper concentrate and supplied it to copper smelters of Japan and South Korea. The acquisition of Magma with its metallurgical plant in San Manuel in Arizona brought BHP to the ranks of fully integrated copper producers with a copper production capacity of copper cathode and copper rod. It was expected that the acquisition would eliminate duplication in productive and management cycles, which consequently could save anywhere from US$ 30 to US$ 40 million per year. At the time of the purchase, the market price of copper was US$ 1.32 per pound. Consolidation of experienced miners and metal makers from Magma and BHP created an efficient copper division, headquartered in San Francisco, California, United States. It was envisaged that the Copper Division would take a dynamic course for further growth of labor productivity through more efficient copper production.

Contacts between BHP and Magma had been rooted in the past. Only 1 year earlier in November 1994, Magma competed the acquisition of the Tintaya copper deposit in Peru. Magma Copper won the tender by offering a more convincing mine development program and paying US$ 269 million to the Government of Peru for the acquisition of 98.43% shares of the Peruvian company Empresa Minera Especial Tintaya S.A. The Magma Company for the next 5 years pledged to spend US$ 85 million on the development of the Tintaya mine. BHP also participated in the tender but assessed Tintaya lower than Magma and so ended up being a loser in this tender. Tintaya skarn sulfide mine extracted ores with 1.75% copper content in the volume of 8,000 tonnes of ore per day and producing about 110 million pounds (50,000 tonnes) of copper per year. The investor undertook to increase the full-time production over the next 2 years, up to 10,000 tonnes of ore per day and engage in the exploitation of the ore with lower copper content. Magma planned to use its SX-EW technology for the processing of copper ore based on the expertise accumulated in Arizona. Magma anticipated that with this technology, it would be able to increase production to 200 million pounds or 90,600 tonnes per year with a production cost at US$ 0.50 per pound.

Before Magma submitted its proposal, it conducted a thorough examination of the state of the prospect at Tintaya. Although the proposal was courageous, yet the implementation of the proposed plan despite certain specific risks was entirely convincing. The company approached the acquisition rather aggressively and took the victory in their hands. Undoubtedly, the outcome affected BHP. This, to some extent, predetermined nature of contacts between the representatives of BHP, represented by its President John Prescott and BHP Minerals Manager Jeremy (Jerry) Ellis with the President of Magma Burgess Winter and Magma Executive Vice President Brad Mills.

The work went quietly, fruitfully, and eventually culminated in the decision of the BHP to purchase Magma Copper, thus becoming the world's second-largest copper producer. In addition to John Prescott, another critical BHP player in the history of the Magma acquisition was Jerry Ellis. Although the top executives of the company took the decision, it was not without a considerable "rock of fortune," or rather "miscalculation." In the year 1996, copper prices collapsed from high US$ 1.30 to US$ 0.86 per pound, while gold prices dropped from US$ 376 to US$ 300 per ounce, and oil from US$ 25 to US$ 15 per barrel.

All these have dramatically worsened the economic performance of the Australian mining giant. Moreover, with the closure of the failed projects listed above, the company had to write off about US$ 8 billion. The amount was very imposing at the time, even for the leading mining company like BHP.

2.10 FIRST STEPS OF JV ERDENET–BHP

Initially, it seemed that apart from renaming of the JV, in which the word "Magma" was to be substituted by "BHP" it would not have a significant impact on the activities of the Erdenet–Magma JV. The same General Director and the same office workers in Ulaanbaatar were retained. However, several new and essential factors would soon arise that would significantly influence further developments.

The geological exploration unit of Magma Copper was a small group of expert exploration geologists. The small exploration office on Santa Rita Road in the suburb of Tucson was purely a work location where rocks, sacks with samples, numerous maps, fragments of drill cores, etc. dominated. The group, which reported to Mark Sander, consisted of several geologists: Eric Seedorff, Chief Geologist, Vernon DeRuyter, managing exploration program in Mexico, Patrick Fahey, head of the exploration program in the Southwest United States, John-Mark Staude, the young geologist who worked with Vernon in Mexico, and finally the Head of North-East Asia Program Sergei Diakov. Several geologists based elsewhere in North or South America were also a part of the team. They were a close-knit group who supported each other and were driven by a great a desire to make a successful mineral discovery.

After the acquisition of Magma by BHP, whose geological exploration unit was several orders of magnitude bigger than Magma's, the geological groups were merged. The Magma explorationists now became part of one of the largest major global mining companies, which were not only world leaders in mineral extraction and production but also in prospecting and mineral exploration.

BHP's global exploration group was headquartered in San Francisco. During that period, the group was led initially by Jerry Ellis, and then by Dick Carter.

The exploration unit of BHP Minerals International Exploration Inc. was headed by a well-known North America geologist, Hugo Dummett. Hugo was relatively new in this post. He took this position from the famous discoverer of the coal basin in Eastern Australia Bowen Basin Oliver Warin who could advance new concepts. Oliver possessed brilliant leadership skills, and this earned him great fame among exploration geologists. Everyone, whoever worked under the direction of Oliver Warin, spoke of him with great respect. Not by accident as his heir, Oliver chose Hugo Dummett, a geologist-explorer, and an enthusiast of new discoveries.

Hugo was a "hands-on" exploration manager always "on the move," traveling from one project to the next. Hugo did not like a remote control. He insisted on having firsthand knowledge of projects before making critical decisions related to them.

All of Hugo's peers and employees remember his extraordinary enthusiasm for the art and science of mineral exploration. He himself was an explorer and discoverer. Like Oliver, he was known as a "people person" with a wonderful sense of humor and a leader who could and would inspire his exploration teams. These are some of Hugo's qualities that will stick forever in the memories of the people who knew the man.

Hugo was instrumental for leading to diamond discoveries in Northwest Territories of Canada. Together with partners, Charles Fipke and Stewart Blusson, he brought BHP to discovery of Ekati diamondiferous kimberlite pipes near Lac de Gras. These discoveries were put into production by BHP in 1998 (Krajick, 2001; Danielsen, 2009).

2.11 LIQUIDATION OF JV ERDENET–BHP

The first complication in the Erdenet–BHP JV arose when funding allocations were requested. Instead, Magma contributed Erdenet's share. It became apparent that Erdenet was unable to contribute their 50% share of the funding. A JV meeting headed by Ts. Sanjii, the General Director of Erdenet and Hugo Dummett, Senior Vice President of BHP, was held in Beijing in early May 1996 in an attempt to find a way for the JV to continue on a 50:50 basis. The partners discussed various options, including financial aid to Erdenet by BHP in the form of a loan. BHP, to save the JV, even offered to change the share formula for the partners proportional to their respective financial contribution (Fig. 2.7).

After numerous fruitless attempts to find a compromise, BHP Minerals management put the project on hold to give Erdenet an opportunity to raise funds. Two options were in the consideration. One was for BHP to resume remote monitoring of exploration activities in Mongolia based out of their Hong Kong office as they had prior acquisition of Magma Copper. The other alternative was to consider conducting exploration in Mongolia on their own.

Five years earlier, in 1991, Mongolia had already enacted the Foreign Investment Law, Government Decree 247, and numerous foreign companies started arriving in the country looking for business opportunities. The mining industry attracted a number of these companies because the Mineral Laws, though not yet finalized, were expected to be attractive to foreign investors.

The General Director of the JV strongly recommended that BHP should continue working in Mongolia as a sole foreign investor. His proposal was to dissolve the JV and register BHP as a legal entity, which would contribute

FIGURE 2.7 Meeting of the JV Erdenet–BHP in Beijing May 1996. Participants (left to right): Peter Leman, Mark Sander, Chris Ford, Don Schissel, Eric Seedorff, Erken Namad, Chris Arndt, Hugo Dummett, Sergei Diakov, Ts. Sanjii, Munkhbaatar, Alex Galimov, and D. Khishigsuren.

100% of the capital. Attempts to restore the Erdenet–BHP JV were still ongoing, but in October 1996, BHP finally exited from the JV.

2.12 BHP MINERALS XXK IN MONGOLIA

The Government Decree 247 in 1991 permitted foreign companies to establish an office in Mongolia with a legal address for a duration of 3 years. During the 3-year period, the companies had to meet a number of conditions set by the Mongolian government. These included a registration fee of US$ 700; setting up an account in the National Bank; a declaration of the amount of foreign funds that were to be invested; the employment of some qualified local staff; and the introduction of modern technology. If these conditions were not met within the 3 years, then the company could apply for a 2-year extension for a fee of US$ 300. The Agency of Foreign Trade would make a decision within 10 days, but in terms of BHP, it was not excessive. The company's head office BHP Minerals International Inc. was in San Francisco; it was the American subsidiary of Australia's BHP, headquartered in Melbourne. The geological exploration unit at that time was within the BHP Minerals International Inc. in San Francisco. Therefore, the procedure of obtaining the approval for the decision to open a representative office in Mongolia and the preparation of the required documentation were not onerous.

The representative office, in general terms, did not have the legal right to conduct business in Mongolia. They were entitled only to explore the possibilities for establishing a business in the country. To conduct mineral exploration, BHP needed to establish a legal entity, which would have the full right to conduct business. It had to be either a registered limited liability company or a branch. BHP applied for the latter.

Two months from the date of the JV Erdenet–BHP liquidation, BHP Minerals filed all the documents required by the Agency for Foreign Investment and Foreign Trade, first for the office registration and then for the BHP Minerals International Exploration Inc. branch, a Delaware company in America. In early August, the registration of the representative office of the company was completed, which entitled BHP to resume its reconnaissance. Sergei Diakov received an appointment to the position of the Manager of the BHP Office, and D. Khishigsuren as the Deputy Manager. In January 1997, we finally received a full registration of the branch as BHP Minerals LLC (or "BHP Минералз XXK" in Mongolian) company.

Following the acquisition of Magma by BHP, the Mongolian program became a part of the Eurasian Region commanded by Donald Schissel. In the region, there were three subregional divisions: European, Far Eastern, and Central Asia. The Central-Asian region included the combined programs in Mongolia, Kazakhstan, and China. Sergei Diakov was appointed as the Manager of the Central Asia subregion reporting to Don Schissel.

In the first half of April 1997, BHP Minerals organized a copper-porphyry seminar in Tucson. Two members of the Central Asian team were invited and S. Sanjdorj from the BHP Mongolian team attended. The seminar was to address various aspects of the search of copper-porphyry deposits, the identification of leached oxidized deposits, diagnostics of limonites as an indicator in areas of secondary sulfide enrichment, identification of clay minerals, and the determination of the level of erosion. An integral part of the seminar was a visit to the Silver Bell deposit, which contained the above ingredients.

The Silver Bell deposit was discovered in 1909 and went into production in 1954. The Asarco Company was mining the deposit. The resource according to USGS was estimated 268 Mt at 0.66% Cu and 0.013% Mo (USGS, Silver Bell). Until 1972, Asarco produced 37.2 million tonnes of ore with copper and molybdenum grades of 0.8% and 0.004%, respectively. On the surface, the deposit was oxidized resulting in the formation of a zone of secondary sulfide enrichment in the form of a chalcocite blanket located above the lower-grade primary ores. The host rocks were quartz granodiorites and quartz monzonites of the Laramides age from 66 to 69 million years. On the surface, the quartz stockwork had pronounced leached cavities filled with limonite and hematite mixed with various types of clays (Titley, 1994). Practical application of distinguishing the typical leach cap minerals required identification of iron oxide and clay minerals in the leach caps.

The technique of defining color mixture of hematite and limonites is helpful for geologists in establishing the presence of chalcocite blankets at depth. The range of clay minerals in the copper porphyry system developed around the porphyry intrusive is normally a mixture of hydrothermal alteration and weathering processes. Kaolin is a common clay mineral in the alteration zones. It occurs both in hydrothermally altered rocks and weathered intermediate and acidic volcanics. The presence of sericite in the clays clearly indicate the hydrothermal nature of kaolinitization. Often in porphyry systems during the late stages, a telescopic overprinting of epithermal processes occurs over the previously deposited porphyry mineralization and associated clays, which results in the formation of another assemblage of clays. Recognizing the type of clays related to porphyry systems is critically important.

For the diagnosis of clay minerals, PIMA equipment was widely used. This is essentially an infrared spectrometer functioning based on measurements of the lengths of the reflected waves from the planes of the crystal lattice of minerals, allowing adequate diagnostics of the clay minerals. The clay products of hydrothermal alteration and weathering are normally very difficult to identify by simple visualization. Based on the spectral analysis in the field, PIMA helped to diagnose clay minerals reliably and quickly. In subsequent years, already under Ivanhoe Mines umbrella, both D. Garamjav and T. Munkhbat went through a similar training with field visits of the porphyry systems in Arizona organized by the University of Arizona in Tucson.

2.13 FOCUS ON SOUTH GOBI

In the new phase of the Mongolian copper program, BHP's priorities switched to copper porphyry systems. In the previous year, the focus of porphyry search in Mongolia was on the Selenga zone in the north and to a lesser extent on southern Mongolia. Reconnaissance during the second field season in the South Gobi revealed the presence of large zones of alteration associated with the eroded Paleozoic volcanic centers. Here wide fields of propylitic and sericitic alteration zones containing zones of silicification were found, all elements of the new model. These eroded systems appeared to have the scope and magnitude of hydrothermal processes that could generate major and economically feasible porphyry deposits. Climatic conditions of limited precipitation in a semi-arid climate are amenable for the development of oxidation zones and associated secondary sulfide enrichment. Clearly, southern Mongolia indicated much greater potential for the discovery of a world-class supergene-enriched copper porphyry deposit.

Chapter 3

Third Field Season—Prospecting for Copper Porphyry Systems

Chapter Outline

Samand Sanjdorj *graduated from the Irkutsk Polytechnic Institute majoring in geophysics in 1973. Until 1997, he worked as an engineer geophysicist in groundwater search division at the Ministry of Water Resources of Mongolia. From January 1997 and onward, he worked as a geologist–contractor in the Mongolian branch of BHP until July 1997, and then he was hired to a permanent position. Sanjdorj was the Oyu Tolgoi field team leader. In 1998, he became Manager of the BHP office in Mongolia, combining fieldwork activities and office work. In 2000, Samand moved to Ivanhoe Mines, where his efforts led to a discovery of groundwater for Oyu Tolgoi. Until recently, Sanjdorj was the Vice President of Oyu Tolgoi; he is currently the company's project manager and Honorary Vice President.*

Discovery of Oyu Tolgoi. https://doi.org/10.1016/B978-0-12-816089-3.00003-2

Sam Carter *was a 1994 graduate of Edinburgh University. After conducting geophysical works in Golder Associates in Ireland, Italy, Turkey, and the United Kingdom, in 1996, Sam was hired by BHP. Sam Carter participated in the design, planning, and execution of geophysical works in Sweden, Kazakhstan, and Mongolia. In 1999, Sam joined Shell Oil Company as a resource engineer. Currently, Sam is still with Shell and continues to work for the Shell Company in the United Kingdom.*

José (Pepe) Perelló*, a graduate of the University of Chile in Santiago, received Masters from Queen's University in Ontario. Pepe participated in geological exploration works on the well-known porphyry copper deposit Escondida in Chile in the company Minera Utah de Chile. Until Mongolia, Pepe was involved in the mapping and exploration of another copper porphyry deposit of BHP, Reko Diq in Pakistan. At present, he is managing the international geological unit of the Chilean mining company Antofagasta Minerals.*

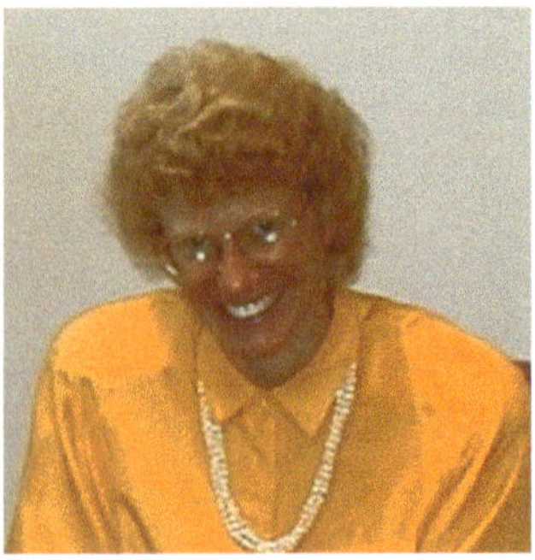

Mary (Theobald) Doherty *has a Bachelor's degree in Geology from the University of Colorado in Boulder and Masters in Geochemistry from the Queens University. From 1984 to 1994, Mary worked at Freeport McMoran as exploration geologist, and for 6 years as Exploration Geochemist in BHP Minerals. She was a Director of Business Development at ALS LLC. Most recently in March 2012, she joined Newmont as Chief Geochemist. Mary combines strong and varied background in geochemistry with exploration geology experience.*

During the period between suspension of the Erdenet-BHP JV in May 1996 and the finalization of the BHP's office registration in August 1996, no fieldwork was conducted in Mongolia. Now that the focus had shifted to exploring for leached copper porphyry systems in the southern Mongolia, preparations began for the next phase of reconnaissance in the Gobi Desert. The plan was initially to revisit previously recommended prospects in the Gobi, such as Shuteen and Ih Shankh, and then to continue exploring other parts of the southern volcanic belt from east to west.

On August 8, 1996, just before field reconnaissance was ready to resume, a cholera epidemic broke out in Mongolia. In the first 3 weeks of the epidemic, 11 people died, and 151 became newly infected. The government officials quarantined the affected parts of Mongolia, and some towns were closed to the public. The epidemic reached its peak in mid-August, and the battle with cholera continued into September. The government decided even to postpone the start of the school year for 2 weeks until September 16. At that point, the field team of D. Cox, D. Garamjav, and T. Munkhbat traveled south to the southern Gobi Desert.

3.1 BHP MINERALS EXPLORATION STRATEGY REVISION

At the same time, the BHP Minerals exploration department was conducting a copper exploration strategy workshop in San Francisco. Representatives from all regions who were involved with copper exploration attended the meeting. Representatives of the production unit of BHP Copper were also invited. At that time, BHP was conducting mineral exploration on a large number of geologically diverse copper properties around the world. The list of copper deposit types included copper porphyry, strata-bound, skarn, and volcanogenic massive sulfide (VMS) deposits. The aim of the workshop was to determine which of these deposit styles had the potential to deliver giant deposits with a high-enough grade that would be highly profitable for BHP. During this time, BHP was experiencing some budget cuts, so the projects had to be prioritized for funding and decisions had to be made around the most efficient exploration methods and tools that could be applied to make discoveries.

After much discussion a consensus emerged that porphyry copper (e.g., Chuquicamata, El Teniente, Escondida, Grasberg, Bingham Canyon, and so forth), strata-bound (e.g., Udokan, Dzhezkazgan, Lake Superior, and so forth), and skarn type styles of deposits (Ok Tedi and Kucing Liar) must be considered. It was agreed that the porphyry copper deposits with enhancement by secondary enrichment, high-grade skarns, or deposits with significantly associated gold mineralization (such as Grasberg) would be given the highest priority. A deposit type that did not make to the list was the VMS type of copper deposit, the exploration program for which were destined for termination. Not an easy decision. Some North American programs were focusing on the search for VMS deposits, and these were all terminated. From the strategical point of view, it was the right decision because the limited funding could go to the prioritized deposit types. As mentioned earlier, the Mongolian program was already in good alignment with his new copper exploration strategy.

After the meeting in San Francisco, Sergei Diakov went directly to Mongolia to join the field team. Upon arrival at the Ulaanbaatar BHP office, Sergei received a satellite telephone call from Dennis Cox informing him that the team had visited a copper occurrence near the village of Khanbogd in the Gobi Desert, which had strong indications of being part of a copper porphyry system.

Early the following morning of September 18, 1996, Sergei and Khishigsuren headed south in their Russian-made UAZ truck. It was a 12-h 600-km-long trip along dusty, ungraveled dirt roads of which only 40 km were paved. This trip was representative of the driving conditions in Mongolia at that time. During dry weather conditions, cruising speeds of 60–70 km per hour could be maintained along sections of the unpaved roads that were in a very good condition. Many of these roads were not wide enough for two vehicles to pass each other safely at speed, but, fortunately, traffic was exceedingly light. However, during the rainy season, these dirt roads would be dangerously slippery with axle-deep muddy patches, and special precautions had to be taken to avoid getting stuck. Consequently, at such locations, drivers would make a fresh pass resulting in a myriad of parallel tracks.

By late afternoon, the lights of the Khanbogd village were coming into view. After some help from the locals, who were already aware of the BHP field camp, Sergei and Khishigsuren arrived at the camp after dark. They found the crew in high spirits huddled around a small campfire sipping tea and discussing plans for the next day.

3.2 KHANBOGD COMPLEX

The copper prospect we were to visit was located in close proximity to the Khanbogd Complex, which is a prominent, circular topographic feature in the southern part of the Gobi Desert, conspicuous on satellite images.

The Khanbogd Complex is the largest, massive, alkaline intrusive rocks in Mongolia occupying an area of 1,500 square kilometers. The complex is Permian in age and represents an intrusion of type An ultraalkaline granitoids enriched in zirconium, niobium, and rare earth elements (Kovalenko et al., 1995). All ultraalkaline complexes manifest both textural and geochemical evidence that hydrothermal fluids transport rare earth elements. The formation of the complex is believed to be either caused by a mantle plume or melting in the asthenosphere from the collision of the Siberian and north-Chinese platforms (Yarmolyuk, Kovalenko, 2002). Such a process causes the formation of an unusual paragenesis of zirconium silicate minerals.

The Khanbogd Complex can be subdivided into an eastern and western half. The Paleozoic age Tsokiot volcanogenic sedimentary complex hosts the ultraalkaline granitoids. The ultraalkaline granites that occur in the western half of the complex have an age date of 292–283 Ma (Kovalenko et al., 2006). This unit is medium- and coarse-grained consisting of 48%–50% quartz, up to

35% potassium feldspar, up to 11% of aegirine, and up to 12% of arfvedsonite. Widespread replacement of arfvedsonite by aegirine forms radial growth structures (Figs. 3.1 and 3.2). Accessory minerals are zirconium silicates, apatite, rutile, and titanite.

Within the Khanbogd Complex, there is also a younger phase of biotite granite dated at 272 Ma (Kovalenko et al., 2006) and numerous circle-shaped ring dikes and radial-shaped dikes.

Upon arrival at the reconnaissance camp in the Khanbogd Complex, Sergei was given an update on the reconnaissance program and, in particular, on the copper showing associated with copper porphyry style alteration. While conducting reconnaissance in the Khanbogd Complex vicinity, the team attempted to revisit some of the copper showings that D. Garamjav had seen on a previous mapping project. During the search, the team found a hill composed

FIGURE 3.1 D. Garamjav in front of the outcrops of Khanbogd Complex holding samples with arfvedsonite.

FIGURE 3.2 Khanbogd Complex. Clusters with arfvedsonite.

of intensely silicified rocks. Here the team noted numerous leached cavities in quartz stockwork veins. According to D. Cox, this outcrop had the appearance of a typical cap rock over the leached porphyry systems.

Early in the morning, the team drove to the hill of silicified rocks about 60 km from the camp. The landscape is a flat sandy and stony desert plain with sparse shrubs and periodic hills, which is typical for the Gobi Desert. Most notable in the area is a table mountain named Javhalant (Fig. 3.3) about 18 km southwest of the silicified hills. There are petroglyphs depicting animals—evidence of ancient nomads who once inhabited this area. Granitoids at the top of Javhalant are dated 324 ± 2 Ma (Gerel et al., 2005).

The silicified hill was nondescript, rising a few tens of meters above the Gobi desert. The hill was about 150 m across and was composed of volcanic rocks of andesitic and andesite-dacitic composition with intense silicification–argillization and some sericite—all saturated with quartz veins. Quartz veins contained numerous voids, indicating the previous presence of sulfides. The ubiquitous distribution of brown limonite and hematite was an evidence for copper sulfide mineralization, which was consequently leached and then moved in a soluble form from the surface to lower horizons. Moreover, the presence of hematite suggested that a chalcocite blanket had been formed by this process and was moved at least several times being ultimately deposited at depth beneath the hill.

Usually a black manganese oxide tarnish covers the stones scattered across the desert, which can conceal evidence of surficial alteration. When viewed with the eye of an expert, the presence of a whitish surficial stain caused by clay and kaolin with limonite could distinguish this hill from other similar mounds in the region. The alteration zone was more noticeable from aerial views (Fig. 3.4) and was subsequently confirmed with Landsat satellite imagery due to an unusual range of colors seen on the images compared to other

FIGURE 3.3 Typical Gobi landscape around Oyu Tolgoi; the table mountain Javhalant is in the background.

FIGURE 3.4 Bird's-eye view of Central Oyu with strongly altered rocks forming a visible color anomaly among the gentle hills in the Gobi. On the right, there is a distinct vegetation anomaly.

unaltered rock outcrops of volcanic rock. Interestingly, on the southeastern end of the hill, at the foot, there was a small clearing under 20 m in diameter covered with green grass growing in the desert, which was a clear sign of the presence of groundwater at a shallow depth. This "birthmark" was very characteristic for this hill and was easily recognizable on all remote-sensing images.

Garamjav found the copper in a small digging by about one-half-meter deep and about 2 m in diameter. According to Garamjav, this small pit indicated a possibility for some ancient copper mining. This copper showing was named Oyu Tolgoi or Turquoise Hill because of the presence of crusts of chrysocolla and malachite. Although we did not find turquoise itself, due to similarity in colors and the abundant presence of magnetite with chlorite, the andesite volcanic rocks looked very dark, almost black. With this background, chrysocolla was conspicuously noticeable, so the name of Oyu Tolgoi was quite appropriate.

The mineralized outcrop in the southwest was named South-West Oyu, and the hill of silicified volcanics was named Central Oyu (Fig. 3.5). However, in west-central Oyu, there was a zone of mineralization different in nature due to numerous thin quartz–magnetite ± pyrite veins within intensely chloritized volcanics. This outcrop was given the name West Oyu (Fig. 3.6). West Oyu, unlike Central Oyu, did not experience any significant leaching, which indicated that the occurrence of copper mineralization had a somewhat different paragenesis.

Traversing to the northeast of Central Oyu revealed that behind the silicified hill of Central Oyu, at a distance of about 300 m, there was a small volcanic hill. Recent clay covers further extension of volcanic rocks to the north. Similar volcanic rocks in this hill also bore signs of weak hydrothermal

(A) (B)

FIGURE 3.5 Quartz stockwork (A) and limonite-hematitic iron oxides (B) at Central Oyu.

(A) (B)

FIGURE 3.6 Copper oxides in the form of chrysocolla (A) at South-West Oyu and outcrops of quartz-magnetite stockwork (B) at West Oyu Tolgoi.

alterations in the form of chlorite—pyritic associations that might be a part of the overall propylitic alteration around the porphyry systems. The hill received a name North Oyu.

Field observations revealed the presence of numerous outcrops with copper mineralization associated with various types of alteration that pointed to the presence of a previously unrecognized porphyry copper system. This was a fascinating development, and it was obvious that this area deserved further attention and exploration to assess its mineral potential.

Furthermore, Oyu Tolgoi hosted several centers with porphyry mineralization, which could coalesce into a major system at depth. Presence of sings of secondary sulfide enrichment raised the spectra of hope that not only the target of scale could be achieved but also that the copper tenor could be greatly enhanced. Although it was somewhat premature to speculate, but perhaps this prospect, found so early in this exploration program, could meet the challenging target goals that BHP was pursuing in Mongolia.

Also noteworthy was that in the Oyu Tolgoi hills, absolutely no evidence of modern exploration had been found. This meant it was a "virgin" unassessed

FIGURE 3.7 D. Garamjav in the process of mapping one of the outcrops of Ih Shankh.

prospect with good exploration prospectivity. No doubt, it was urgent to "stake" the ground and apply for an exploration license.

After visiting this prospect and recording the presence of leached stockwork there and the widespread distribution of porphyry style alteration, it became clear that Oyu Tolgoi has a potential to become a leading copper project among all other copper prospects we had seen in Mongolia. Oyu Tolgoi seemed to have the key attributes of a promising porphyry-type system that the previous Erdenet-Magma JV and then BHP had been seeking.

It was now urgent that BHP apply for an exploration license. To delineate the license boundaries, a quick reconnaissance around Oyu Tolgoi was conducted. To the north, at Shivee Ovoo, we discovered altered volcanics. After visiting several outcrops to the south, the limits of the license area were determined with a good degree of confidence.

In addition to Oyu Tolgoi, another prospect in the Southern Gobi, Ih Shankh, (Fig. 3.7) also demonstrated quite encouraging alteration features. Based on the reconnaissance results, Ih Shankh contained several zones of intense argillic alteration with the development of alunite and hematite, accompanied by noticeable quartz stockworks and occasional individual quartz veins.

3.3 APPLICATION FOR EXPLORATION LICENSES

Immediately upon returning to Ulaanbaatar, S. Diakov, on behalf of the Representative Office of BHP, submitted the applications for exploration licenses for Oyu Tolgoi and Ih Shankh at Mongolia's mineral resources agency. The coordinates for the corners of the Oyu Tolgoi license were defined as N42°47', E106°30', N43°08', E107°00′, which encompassed an area of 1,350 km^2. Ih Shankh was delimited by coordinates N43°30', E105°45', N43°45', E106°15′ covering an area of 1,168.5 km^2 (Fig. 3.8). Both license areas were located in South Gobi Aimak, within the administrative center of Dalanzadgad (Fig. 3.9).

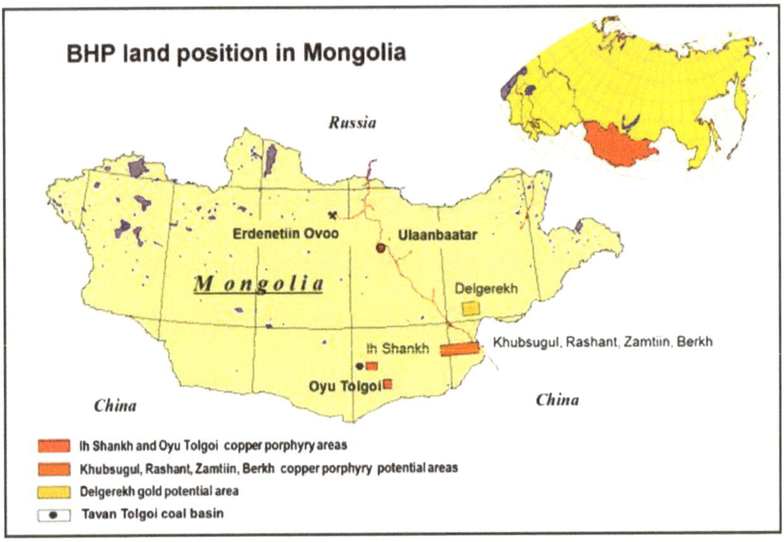

FIGURE 3.8 Map of Mongolia with license areas of Oyu Tolgoi, Ih Shankh, and other regions recommended in South Gobi.

FIGURE 3.9 Dalanzadgad, capital of South Gobi Aimak.

In January 1997, BHP completed registration of its branch (registration no.: 2078422) and then, shortly after that on February 17, BHP also received exploration licenses for Oyu Tolgoi and Ih Shankh. Each exploration license was valid for 3 years.

It was very satisfying for our team that through systematic, traditional "boot and hammer" style mineral exploration and by rigorously adhering to our exploration model, we were able to identify and acquire such high-quality properties in a relatively short period of time. This could only be achieved by an experienced, well-rounded team that has the expertise to recognize a copper porphyry setting that offers size potential and the likelihood of supergene enrichment and is able to identify the critical, but often subtle, geological criteria in the field that are required by the model.

3.4 PREPARATION FOR DETAILED PROPERTY EXPLORATION

In 1997, activities during the field season were primarily focused on determination of the prospectivity of Oyu Tolgoi and secondarily Ih Shankh. Preparation for the field season began in March.

It soon became apparent that very little information was available for the prospect areas of interest. There were no adequate topographic maps, and geological, geophysical, and geochemical data were available only at regional scale. To assess the prospectivity of Oyu Tolgoi, a detailed geological map had to be produced over the project area. Owing to the lack of adequate topographic maps, the first field program for the 1997 season was to complete a topographic survey. This topographic map would then be used as a base for the geological, geophysical, and geochemical data.

The sequence of field program implementation was as follows: (1) the topographic survey, (2) geophysical surveys which could be performed before all the snow had melted, (3) geological mapping, and (4) a geochemical survey. After compiling the various data sets, anomalies would be prioritized for exploration drill testing. Remote sensing was the first step in preparing for field mapping.

3.5 REMOTE SENSING

Remote sensing was the first step in preparing for field mapping. The remote data were generated by a thematic (TM) survey from Landsat satellites. A TM multispectral survey was conducted by scanners in seven channels within visible and infrared parts of the spectrum (Fig. 3.10).

Bedrock exposure on the Oyu Tolgoi license was relatively abundant, and there was very little vegetation cover. These were ideal conditions for the use of remote-sensing methods. In addition, to highlighting the presence of altered rocks, the processed Landsat images also provided significant information on structures, intrusions, and exposed rocks as opposed to cover and infrastructure.

An idealized model of porphyry copper deposits consists of a central zone of host rocks containing quartz veinlets with iron oxides, surrounded by a halo of hydrothermally altered host rocks, dominated with clay minerals. The central zone is usually significantly smaller in comparison to the area of alteration around it. Therefore, in remote sensing, the primary objective is to determine the area of altered rocks. Clay minerals are usually easy to detect,

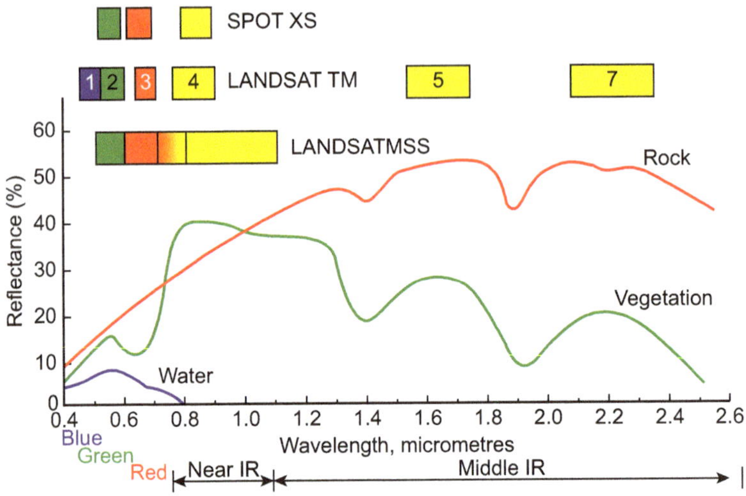

FIGURE 3.10 Ranges of frequencies covered by Landsat TM images. *IR*, infrared; *MSS*, multispectral scanner system; *TM*, thematic.

but identification of individual zones with specific sets of clay minerals requires images with higher resolution.

Weathering influences the effects seen in the photos. Even a small crust of weathered rocks of some 50 μm in thickness may significantly affect the reflection range. Another complication is that the clay minerals may be either of hydrothermal origin or just a result of simple weathering. Therefore, false anomalies can easily appear on the images meaning that not all clay minerals, showing on the Landsat TM photos, are necessarily of a hydrothermal origin. Specific features of the distribution of altered minerals give some hint in the contrast of weathered clays from those caused by hydrothermal alterations. Hence, to adequately determine whether clays originated in hydrothermal or in the weathering processes, one must look into the morphology and distribution of clay minerals.

To identify the areas of alteration within the licensed ground of Oyu Tolgoi and Ih Shankh, three images were acquired. One of the images was a black and white SPOT photo taken on August 29, 1996, which had better resolution than Landsat. The other two images were snapshots of seven channel Landsat TM. Pictures of TM had resolution 28.5 m × 28.5 m, which made it possible to produce the interpretation at a scale of 1:100,000. However, SPOT images had a higher resolution of 10 m × 10 m; therefore, they were more suitable for creating interpretation maps of a larger scale of 1:33,000. The combination of Landsat TM pictures and SPOT images usually significantly improve the resolution of multispectral spectral images based on TM (Fig. 3.11).

For Oyu Tolgoi, a combination of Landsat TM and SPOT images were used. Owing to the details of the topographic relief of various geological

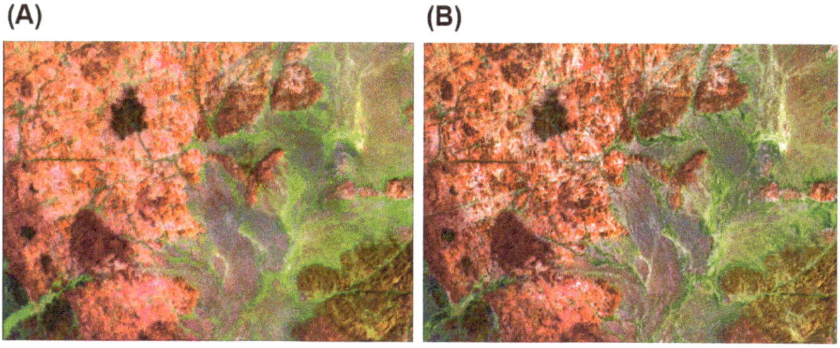

FIGURE 3.11 TM images without SPOT (A) and with SPOT (B).

features, structural elements within the superimposed images appeared to be clearer and sharper. These images also proved to be a useful tool in the hands of the geological mappers.

In preparation for fieldwork, both Landsat and SPOT satellite images were processed and analyzed. Spectral analysis in various combinations of red-green-blue (RGB) channels showed that Hill Central Oyu was easily identified by magenta color in pictures 741 RGB.

Within the Oyu Tolgoi and Ih Shankh licenses, Landsat imagery was used to identify outcrops, which then were field-checked during the summer. On the Landsat images, the Tolgoi Oyu hills stood out as visible anomalous areas of hydrothermally altered rocks. The Central Oyu Hill was particularly distinguishable. The hills of Ih Shankh revealed several similar zones, and all of these sites were earmarked for field follow-up.

Two images in the RGB spectrum were generated at frequencies of 321 and 741. A snapshot of 741 was taken at a height of 705 km in both visible and invisible infrared frequencies. Red color reflects the presence of iron oxides, green indicates vegetation cover, and blue indicates areas with a high content of clay minerals. The combination of the clay alteration and iron oxides on the images was marked in the purple color (Fig. 3.12).

These aforementioned sites represented the most exciting areas for further study because they could indicate cap rocks of leached zones over porphyry systems.

Based on the interpretation of these remote images, satellite photos in three scales 1:100,000, 1:50,000 and 1:25,000 were generated for the geological mappers. In these pictures, bright red highlighted the zones of altered rocks, and gray–white showed the rest of the rocks (Fig. 3.13).

The acquisition of the satellite images was followed by a compilation of geological, geophysical data, and structural maps. All this information was stored in the database in ArcView format, which the Oyu Tolgoi team started to accumulate.

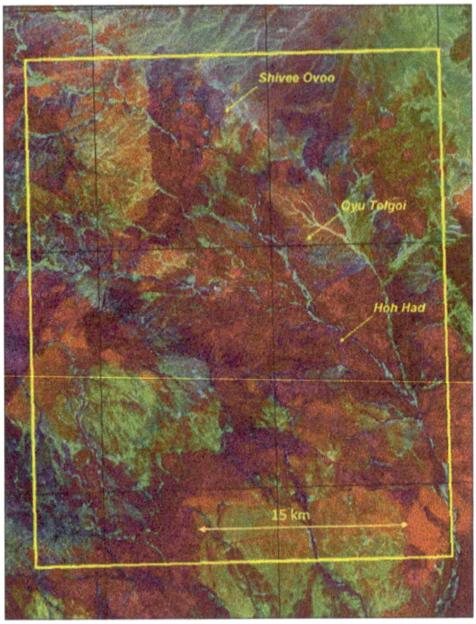

FIGURE 3.12 Landsat image 741 RGB. Zones of alteration are purple. The yellow rectangle shows the boundary of the exploration license.

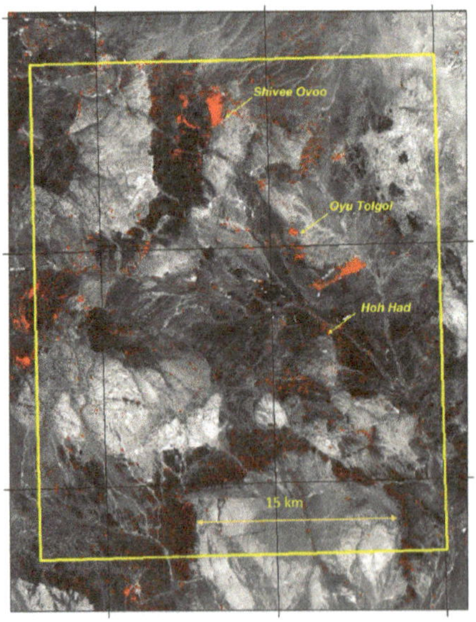

FIGURE 3.13 Landsat image with zones of alteration in red color on a gray background.

3.6 MAPPING, GEOPHYSICAL, AND GEOCHEMICAL SURVEYS

To generate a topographic map, we contracted MonMap, a local engineering service company, who had the capability of conducting topographic and geophysical surveys.

The principal selected geophysical methods were ground magnetics and transient electromagnetics (TEM). The magnetic survey was selected primarily to delineate magnetic lows, which would potentially reflect areas of magnetic destruction associated with the zones of phyllic alteration. The magnetics could also detect magnetically active zones, which could identify potassic alteration, as well as magnetite-bearing copper skarns beneath the hill of Central Oyu. The purpose of the TEM method was to delineate the distribution of any zones of secondary sulfide enrichment, not only under the cap rock of leached stockwork but also under the cover of overlying younger rocks surrounding the Oyu Tolgoi hills. From the outset, the issue of the thickness of the Cretaceous- and Tertiary-age sediments around the Oyu Tolgoi was in consideration. For this, the method of vertical electric sounding (VES) was selected.

Attempts to find a suitable local contractor who could perform the necessary level of TEM work, unfortunately, did not yield any fruitful results, and we consequently resorted to the alternative method of gradient-induced polarization (IP), which would record the presence of primary disseminated sulfide mineralization at depth. The logic was that if the IP anomaly was associated with signs of surface in the central zone, then the secondary sulfide enrichment with chalcocite blanket underneath Central Oyu would merely have to be found there above the zones with primary mineralization.

The sequence of fieldwork was determined by the weather and climatic conditions in the area, beginning with the topographic survey first, which was followed by geophysics as topographic base was laid out. With the spring weather in the Gobi warming up, the geological mapping and subsequently geochemical survey were initiated. All these activities were to help us in delineation of anomalies for exploration drill testing.

Howard Golden, Regional Chief Geophysicist at the London BHP Office, supervised geophysical work. Geophysica Service Company, a local private company, was selected to perform the IP survey (Fig. 3.14). Sam Carter, a young geophysicist from the BHP Minerals exploration office in London, was seconded to oversee the contracted geophysical work in the field.

The works mentioned previously allowed creating topographic, magnetic, and IP maps at a scale of 1:25,000. Geological mapping began in late May, without waiting for the full completion of topographic and geophysical surveys. There was a particular reason for this. Owing to a limited field season, we had a short time to check the recommended anomalies by drilling before the onset of winter.

(A) (B)

FIGURE 3.14 MonMap and Geophysica Service equipment at Oyu Tolgoi, including the magnetometer (A) and the IP equipment (B).

In principle, it was the right approach. Together with the mapping, we also planned to conduct a systematic geochemical sampling program, which required some time for all geochemical steps, including sample collection; sample preparation; their analysis, processing, and interpreting the geochemical information; and then finally generating recommendations based on the interpreted results. In general, the timetable was laid out in such a way that at the end of the year, on the basis of the data collected from geological-geophysical-geochemical observations, we would have all necessary information enabling us to conclude first impressions of what Oyu Tolgoi geologically was and what its economic potential could be. It was also needed to determine the follow-up actions to be undertaken further to assess better the mineralization found there.

We must say that the geophysical work with topographic survey began around late March and early April, and they went on in quite severe conditions. Night temperatures occasionally dropped to subzero and below the freezing point. As always, during the powerful springtime, winds exacerbate the weather in Mongolia. Geophysicists set up their local gher (tent) at the foot of the Central Oyu hill downwind (Fig. 3.15).

FIGURE 3.15 First field camp at Oyu Tolgoi installed by Geophysica Service.

FIGURE 3.16 Laying out cables for IP profiling.

Sam and Sanjdorj checked all geophysical field measurements on the spot, and then they transmitted the data via e-mail over the internet through a satellite connection with the London BHP Office. The topographic and geophysical work began on April 13 and was completed on May 15, 1997 (Fig. 3.16). The magnetic survey, which was conducted simultaneously with the topographic survey, covered a 7 km × 8 km area within the Oyu Tolgoi hills. Both the magnetic and IP data were collected on a 250 m × 50 m grid, but the latter survey only covered a 3 km × 4 km rectangle around the same center. The array of IP lines had south–north orientation. After it became apparent that IP anomalies continued to the south, the IP survey area was also extended by 2 km farther in this direction.

3.6.1 Topographic Survey

The MonMap Company, which was contracted to conduct the topographic survey, laid out the base of the surveyed ground using Swiss theodolite Leica. From geodetic points, auxiliary anchor points were made, which were then formed into eight basic profile lines in the latitudinal direction, as well as two connecting profile lines in the meridional direction. Profile lines were located at a distance of 250 m from each other, and measurement points along the profile lines were marked at a distance of 50 m. All coordinates and directions were referenced to the Pulkovo system using UTM48 projection. Each point on the ground was pegged with markers indicating the coordinates of the point on the marker label (Fig. 3.17).

The survey area was quite flat with maximum relief of 60 m (Fig. 3.18). Oyu Tolgoi Hill, with coordinates 650996E and 4764034N (UTM48), was in the center of the survey area.

3.6.2 Results of Geophysical Survey

For the ground magnetic survey, the contractor MonMap Company used three Scintrex ENVI-MAG magnetometers. The method used by MonMap for the

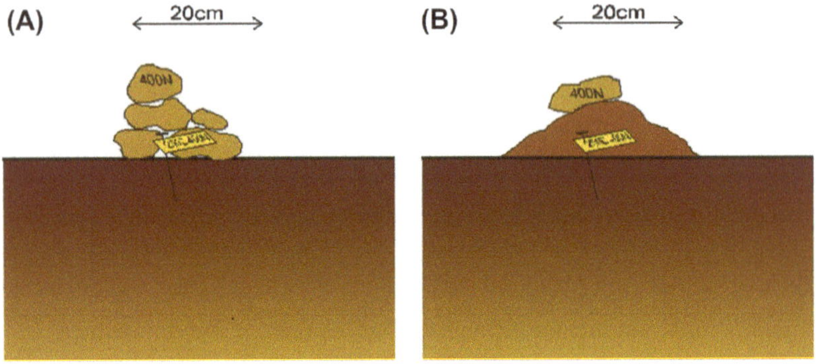

FIGURE 3.17 Two types of local grid peg markers: pile of stones (A) and pile of soil (B).

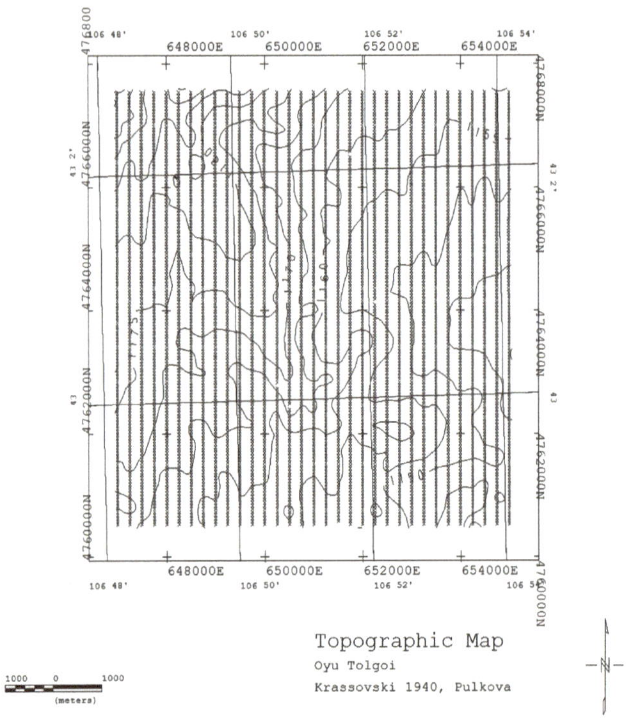

FIGURE 3.18 Map of profile lines for topographic and geophysical survey.

survey was multiinstrumental. One of the magnetometers was used as a base station to record changes in the magnetic field, and the other two were used as mobile stations for measuring the magnetic field at the station sites. Care had to be taken during the magnetic survey to avoid interference from currents generated by the IP survey. Interference was noticeable when the surveys were within about 1 km of each other.

Magnetic data capturing was smoothed out by a cell of 75 m. Then the magnetic data were reduced to a pole for positioning the magnetic anomaly directly over the magnetic body that was causing this anomaly. During the survey, the geophysicists also took magnetic susceptibility readings of the host and overlying rocks. The magnetic susceptibility data served to enhance the interpretation of the magnetic survey results. It was determined that eolian sands, diorite dikes, and basalt units had elevated magnetic susceptibility values due to the presence of magnetite.

The magnetic data proved to be useful in highlighting geological structures and in some cases the bedrock geology. The north–northwest and the north–south structural trends were quite distinct. However, the northwest-trending structures were less pronounced. Central Oyu Tolgoi was located at the cross junction of these structures.

The geophysical results were plotted on 1:25,000 scale maps. All anomalous areas were then identified and highlighted for further follow-up. The magnetic data showed that Oyu Tolgoi was situated on the southern edge of a 3 square kilometers negative magnetic anomaly (Fig. 3.19). This magnetic low is spatially associated with widespread, altered rock exposures found during the reconnaissance program. It is very probable that the result of the magnetic destruction caused by a zone of phyllic, argillic, and/or propylitic alteration is likely a part of the Oyu porphyry system. South Oyu and South-West Oyu on

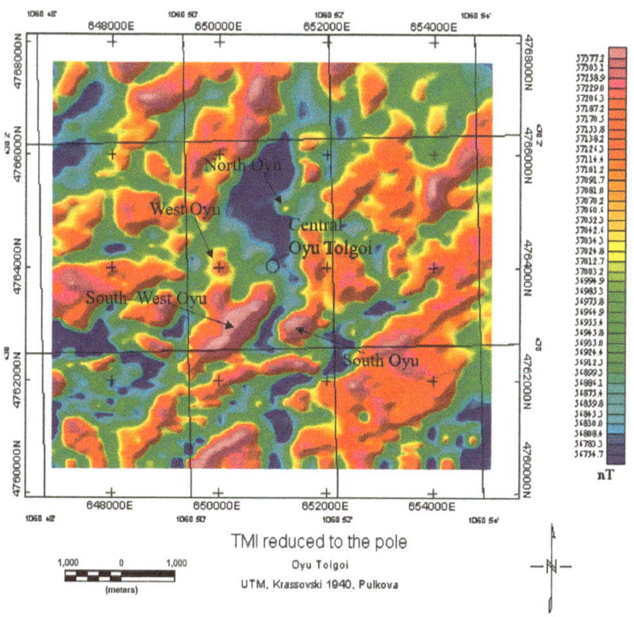

FIGURE 3.19 Map of intensity of magnetic fields at Oyu Tolgoi. *TMI*, Total Magnetic Intensity.

the other hand came out as pronounced magnetic anomalies. Both were explained by the presence of magnetite in the altered rocks at these prospects.

The Geophysica Service using a modified version of the equipment performed two profiles. One profile was laid out in the east-west direction north of the Central Oyu Hill along the lines of the Oyu Tolgoi 750N. The other profile was put in a south-north direction through 250W. The results of the magnetic survey were processed, and a general map of the magnetic field reduced to pole was generated (Fig. 3.20).

The IP survey was conducted with a modified version of an EAP ATS-3000 time domain analog IP system, and apparent resistivity was recorded with a 4.5-kW Honda generator. Adjustments were made to the equipment to separate the transmitter from the receiver, which reduced the need for long cables, hence minimizing the adverse effects of telluric noise. The IP transmitter put a 5 A current at 250 V into the ground in rectangular waves. The power was then turned off, and half a second later, the induced voltage in the ground was measured. The ratio of the measured voltage was calculated as a percentage of the IP. However, creating a favorable ratio between the voltages put into the ground and the measured values proved to be problematic. The upper layers of clay and salty soil near the subsurface prevented the current from reaching the bedrock. Therefore each measurement had to be carefully monitored and controlled. Sanjdorj successfully coped with this task. Later on, a similar approach was introduced in the searches for groundwater in Mongolia.

Given that the maximum distance between the dipoles of point A and point B (Fig. 3.21) was at 2,200 m, the penetration depth was estimated to be about 220 m from the surface. The measured dipole length was chosen to be 200 m (points M and N), and the measured distance was 50 m.

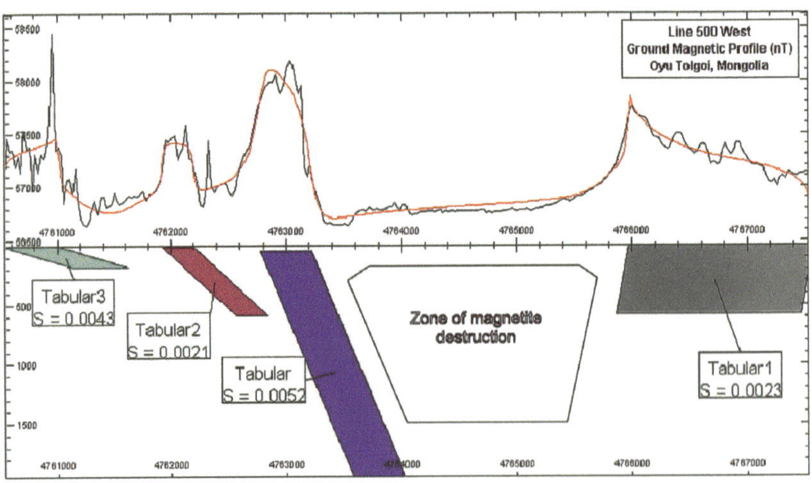

FIGURE 3.20 Geophysical profile across Oyu Tolgoi.

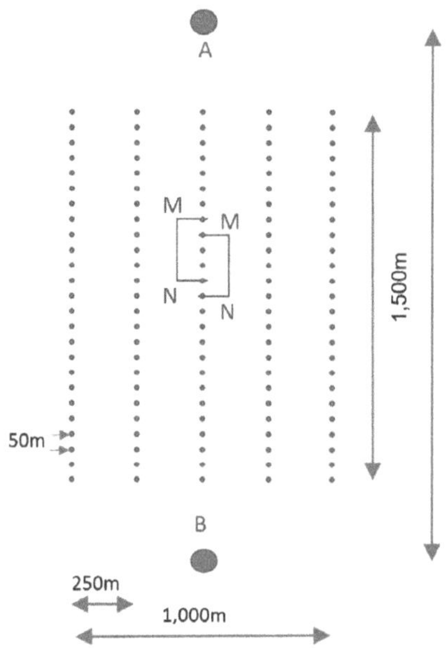

FIGURE 3.21 Plan of layout stations for IP survey. *IP*, induced polarization.

Central Oyu Hill produced a 1 km × 0.8 km IP anomaly with up to 15% chargeability (Fig. 3.22). According to the local geophysicists, the strength of the chargeability was unprecedented in their survey experience in Mongolia. Such a value would be expected to be indicative of sulfides at depth, including copper sulfides, because Central Oyu is known to be a copper showing. The ultimate hope was that much of the sulphide would be in the form of chalcocite, and then Oyu Tolgoi could be a supergene enriched copper porphyry. Of course, there was always a possibility that the sulphide was merely barren pyrite. Although the IP anomaly did not have a strong coincidence with the magnetic anomaly, there was some overlap, it was becoming clear that the data indicated that Central Oyu represented an anomaly confirmed by several methods that ultimately had to be resolved by drilling. If the thickness of the chalcocite blanket was 50 m, then the enriched mineralization of this size could produce in the order of 100 million tonnes of ore. Although quite optimistic, the thought of this possibility was inspiring.

Interpretation of VES conductivity data indicated that the groundwater table was covered by a 5-m thick layer of recent sediments, with resistance readings ranging from 16 to 180 Ω m (Fig. 3.23). Below the groundwater was a blanket of 20–30 m thickness with values of 7–9 Ω m, apparently reflecting a layer of mudstone with sulfides. Below, the survey showed a body of higher resistivity 220 Ω m. To the west, the calculations indicated a body with a very high resistivity at depth, which could not be explained in the field.

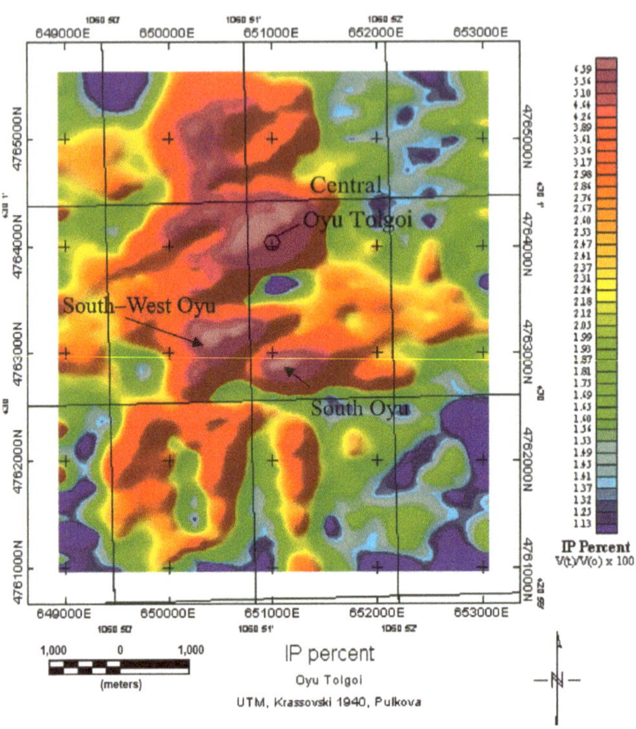

FIGURE 3.22 Map of induced polarization at Oyu Tolgoi. *IP*, induced polarization.

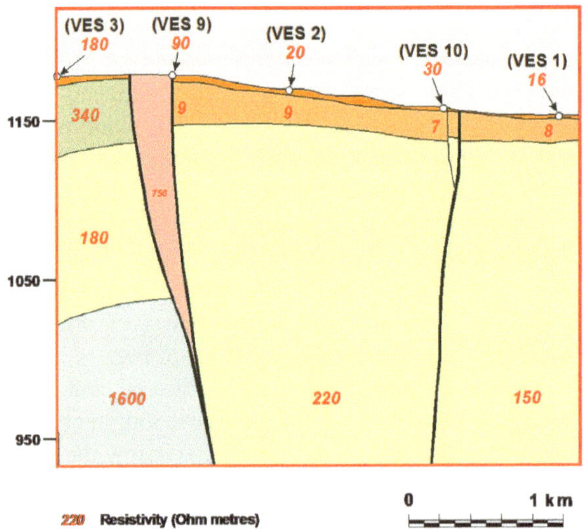

FIGURE 3.23 Profile of VES with interpretation of measured resistivity. *VES*, vertical electric sounding.

Based on geophysical results, several geological models appeared to fit into the known Oyu Tolgoi geological framework namely:

- Oyu Tolgoi was a porphyry system without potassic core. The IP anomaly was generated by sulfides in the phyllic zone or, less likely, by sulphide-bearing magnetic iron skarns in the south reflected by the strong magnetics.
- Oyu Tolgoi was a porphyry system with a magnetite–potassic core. The potassic zone is defined by the presence of magnetite with associated silicification, phyllitization, and propylitization, determining the distribution of magnetite. Sulfides caused the IP anomaly in the phyllic and potassic alteration zones.

Without significant preference to any of the aforementioned options, geophysicists recommended drilling five drill holes to confirm the results and confirmation from the ongoing geological mapping and geochemical survey (Table 3.1). Geophysicists agreed that beneath Central Oyu, there should be a chalcocite blanket present. To confirm this interpretation, they recommended drilling there at least one of the planned boreholes. In case of confirmation of the presence of a chalcocite blanket, geophysicists recommended considering airborne TEM methods for future exploration.

3.7 RESULTS OF GEOLOGIC MAPPING

By the end of May 1997, the field camp at Oyu Tolgoi was expanded to accommodate a larger field crew of about 12 members to conduct geological mapping and geochemical surveys. Now there were six ghers at the camp (Fig. 3.24). The camp site was carefully selected to be situated at a central location within the license area to minimize travel and environmental impact on the ground. A minimum number of roads were made to access the survey areas, and even the driveways and footpaths within the camp area were carefully planned to minimize the impact of traffic. This was performed in consideration of the fragile desert surface that requires a long time to "heal" because of the arid climate. The drivers and crew were expected to adhere to these tracks and paths as strictly as possible at all times.

As mentioned previously, Sanjdorj participated in the seminar on copper porphyry deposits organized by BHP Copper in Tucson in April, 1996. This seminar was mainly focused on studying and mapping leached porphyry systems. Field visits were made to Morenci, Silver Bell, and San Manuel in Arizona. The new knowledge that Sanjdorj brought back to Mongolia and applied at Oyu Tolgoi was invaluable to the project.

BHP's Oyu Tolgoi 1:25,000 scale geological mapping program commenced in May, 1997. The map base was taken from E. Burenhuu's 1:200,000 scale government-funded geological map produced in 1995 and from SPOT and Landsat images. The mapping team, under the leadership of D. Cox, consisted of D. Garamjav, S. Sanjdorj, T. Munkhbat, and G. Oyun. The team mapped,

TABLE 3.1 List of Geophysical Anomalies Recommended for Drill Testing

Rank	Location	Nearest Peg	Target Depth	Character	Prediction
1	651000 E, 4764085 N	Line 0 50 N	VES = 40 m	Magnetite destruction highest IP (15%)	Ore or pyrite shell
2	650500 E, 4763135 N	Line 500 W 900 S	VES = 40 m Mag < 50	Highest mag High IP (7%)	Mag core or mag skarn
3	650750 E, 4765035 N	Line 250 W 1000 N	VES = 40 m	Center of mag destruction and 1,300 m wide > 4% IP zone	Low-grade sulphide potassic core
4	650250 E, 4765085 N	Line 750 W 1050 N	VES = 20 m	Magnetite destruction local high IP (6%)	Ore or pyrite shell
6	651250 E, 4762885 N	Line 250 W 1150 N	VES = 30 m Mag < 99 m	High mag high IP (6.5%)	Mag core or mag skarn

Mag, magnetic; *VES*, vertical electric sounding.

FIGURE 3.24 Field camp of BHP. Management of BHP Minerals Exploration Inc. visiting the project in August 1997.

documented, and sampled all main exposures in the Oyu Tolgoi area, including North Oyu, Central Oyu, South-West Oyu, and West Oyu. Within the remainder of the license area, the reconnaissance survey focused primarily on field-checking the Landsat color anomalies. Observations around Shivee Ovoo and Khoh Khad confirmed that these prospects deserved further attention. Based on the preliminary mapping results, it was recommended to conduct sets of geophysical surveys identical to those at Oyu Tolgoi prospect.

By mid-summer, Jose (Pepe) Perelló, a senior BHP geologist, joined the mapping team. He brought extensive experience gained at BHP's Escondida and Reko Diq copper porphyry deposits located in Chile and Pakistan, respectively. More specifically, Pepe was experienced in the study of relic sulphide textures in leached caps and the mapping and core logging of copper porphyry systems. He was also instrumental in improving the efficacies in the process of interpreting and documenting the salient geologic data.

3.7.1 Lithology and Structure

The Carboniferous-age volcanogenic sedimentary sequences were considered to be the oldest rocks within the licensed area. Sandstone, containing small crystals of sanidine and intercalations of tuff, is the basal unit. It is overlain by layers of lavas, andesite + dacite tuffs, plunging at a shallow angle to the north, and basalts with phenocrysts of clinopyroxene and olivine, which is in turn overlain by a sequence of andesite-dacitic lavas. None of the rock units in this area had ever been dated for age, and it had always been assumed that the overlying volcanics were Devonian (Burenhuu, 1995) based on the lithological similarity to the volcanic rocks from other areas where they had been dated. As a result, the underlying sandstones were assumed to be Silurian. During the course of mapping, the remains of fauna were found on the licensed area, and these were submitted to Prof. Nan Arens of the Museum of Paleontology at the

University of California, Berkeley. The fauna were dated to be Carboniferous, and Prof. Arens believed them to be the remnants of cicadas of the late Carboniferous period. Based on this new age date, it may be assumed that all rocks on the license area that are older than the Permian-age Khanbogd Complex are of Carboniferous age.

Central Oyu is about 20 m higher in relief than the surrounding terrain. At first glance, it blended with the surrounding rocky desert landscape. However, from a helicopter and from Landsat images, it was evident that Central Oyu was significantly lighter in tone from the surrounding hills. This was due to intense hydrothermal alteration of the host rocks. The intensity of the alteration was such that it was impossible to determine the protolith; hence it was difficult to establish the original nature of the rock. However, on the northeastern slope, phenocrysts of feldspar were found to be preserved in the groundmass of sericite and clay minerals. This rock unit was then diagnosed to be feldspar porphyry intrusion. A few pale-colored aplite dikes are situated on the southern slope of Central Oyu Hill. They did not show any signs of hydrothermal alteration appearing to be postmineral. Their presence indicated that the porphyry system was dynamic, developing through several stages of magmatism and related hydrothermal activity.

The majority of volcanic rocks in Oyu Tolgoi area have dacitic, andesitic, and rhyolitic compositions. The absence of signs of lava flows or sedimentary lamination in the volcanic rocks prevented accurate determination of their bedding. For the most part, hydrothermal alteration was too intense to preserve any primary lithological characteristics. Relatively unaltered rocks could only be found in the northeastern part of the license area. These volcanic rocks occur between several outcrops of massive granite plutons. As a result, these volcanic rocks are metamorphosed to amphibolite facies and are cross-cut by chloritized and biotitized faults (Fig. 3.25).

FIGURE 3.25 View on Central Oyu Tolgoi Hill from south.

The dacites tend to be light gray–colored rocks with phenocrysts of sanidine, plagioclase, biotite, and hornblende. Locally rounded 10-cm structures suggest possible subaqueous deposition. Andesites are gray to dark gray rocks with phenocrysts of plagioclase and hornblende. Rhyolites are light-colored rocks with fine crystalline textures, with phenocrysts of quartz and sanidine.

A system of tectonic faults with a dominant northwestern and northeastern orientation, as well as sublongitudinal and sublatitudinal orientations, dissected the licensed area.

The outcrops of Central Oyu Tolgoi, as mentioned previously, were intensely affected by phyllic and argillic alteration associated with silicification. The dominant clay mineral was illite, which in the course of weathering turned into kaolin. At Central Oyu Tolgoi, advanced argillic alteration containing alunite, dickite, kaolinite, pyrophyllite, and zunyite was also developed. The zone of advanced argillic alteration extends to the northeastern slope, and the phyllic alteration is confined to the southern slope. Overall, the dominant clay mineral was illite, which during the course of weathering, turned into kaolinite. Locally, veinlets of supergene alunite, from 1 to 3 mm in thickness, were found on fresh surfaces. Here the alunite resembles porcelain. On the western slope of Central Oyu, within the zone of well-developed advanced argillic alteration, narrow breccia dikes were found. The dikes were up to 50-cm wide and traceable along the strike for lengths of 40 m or more. The fragments of silicified volcanic rocks and quartz material in breccia were isometric in shape from 0.5 to 5 cm in diameter and were predominantly angular and only occasionally subrounded.

At Central Oyu Tolgoi, quartz veins and veinlets with cavities, locally filled with limonite and hematite, were found in great abundance. The frequency of the veinlets ranged from 10 to 50 per square meter, reaching the highest density in the zone of advanced argillization where the veins coalesce into stockworks (Fig. 3.26). The orientation of the veinlets here is random with angles of dip ranging from subhorizontal to steeply vertical. Veinlets of less than 1 mm to 1 cm thickness and rarely up to 10-cm thick were ubiquitous across the stockwork. On the western slope, in the advanced argillic zone, zones of massive silicification up to 1-m thick and more than 10-m long occur. The host rocks were interpreted to be silicified volcanics.

3.7.2 Geochemistry of Host and Cover Rocks

According to the plan of 1997 fieldwork activities, we collected regular geochemical samples, including rock chip samples from bedrock in the outcrops, soil samples from overlying sediments, and also soil BLEG samples along temporary water streams. The geochemical samples were first sent to the Ana Labs preparation facility in Ulaanbaatar for crushing, grinding, and reduction. From there, the samples were shipped to the Chemex laboratory in

FIGURE 3.26 Quartz stockworking in the western slope of the Central Oyu Tolgoi Hill in the advanced argillic altered rocks.

Vancouver for Induced Coupled Plasma (ICP) analysis. The Bulk Leachable Extractable Gold (BLEG) samples collected along the rivers and temporary watercourses were dispatched for analysis to Australia.

Mary Doherty, BHP exploration geochemist, supervised all geochemical work at Oyu Tolgoi from 1997 to 1998. Geologists involved in mapping, mostly S. Sanjdorj, were also helping Mary in collecting the geochemical samples. A broad circle of geologists and geochemists from BHP discussed the results of geochemical analysis, including a well-known Canadian geochemistry Bill Coker.

Soil geochemistry was considered at Oyu Tolgoi first in an attempt to trace the extension of known mineralization under areas where it was concealed by postmineral cover rocks and second where geophysical anomalies had not been resolved. The approach was to conduct an orientation survey outward from the exposed mineralizations at Central, South, and North Oyu Tolgoi to determine if mineralization under postmineral cover rocks could be traced with soil sampling in this environment. Samples were taken along five lines. Another component of the orientation was to determine an optimal sampling interval. This was done by taking samples at various intervals.

The strongest anomalies reflected the extensions of the known mineralization at Central, South, and North Oyu Tolgoi. Taking into account the differences in style and mineral assemblages in these prospects, some variations in their secondary halos were also manifested. The Central Oyu prospect yielded an intense copper-molybdenum-arsenic-vanadium anomaly, whereas South Oyu yielded a strong copper anomaly with some elevated values of molybdenum, lead, zinc, and weaker gold.

Based on the positive results of these orientation geochemical surveys, it was concluded that the soil geochemistry in the Oyu Tolgoi environment should be an effective tool for discovery of mineralization hidden under the postmineral overburden. Consequently, a more regional program was undertaken during 1997 and 1998. Overall, the regional survey confirmed that Oyu Tolgoi has elevated contents of copper-arsenic-gold-molybdenum and, to a lesser extent, elevated lead, zinc, gold, and silver. The soil program helped to outline large geochemical anomalies to the south and east of Central Oyu Tolgoi. Here the presence of anomalous copper and related elements pointed to a possible continuation of the mineralization under the cover in the southeastern and northeastern direction. Consequently, a proposal was made to drill test these trends.

The second soil anomaly was ring shaped consisting of elevated sodium-calcium-strontium-barium with a low value of iron. The significance of these results is that calcium and strontium usually develop along the tectonic fault zones, and a lower iron content is typically found over deposits where a transition of iron ions from Fe^{2+} to Fe^{3+} occurs. However, the consensus was that further experimental work needed to be performed to confirm this conclusion.

Pepe Perelló recommended conducting a special study of relict sulfides in an attempt to better characterize the copper mineralization in the hypogene zone at Central Oyu. The study involves identifying the types of hypogene minerals, such as chalcopyrite and bornite, which are "sealed" in the quartz veins and to determine the copper grade of these quartz veins. During the process of formation of a copper porphyry deposit, copper sulfide is disseminated both in the host porphyry rock and in the stockwork quartz veins. These accumulations of copper sulfides "sealed" in quartz during hypogene mineralization remained in situ and intact due to their isolation from exposure to oxygen and water oxidation. The study of relict sulfides in leached zones proved to be extremely useful, as it allowed determining the mineralogical composition of the primary ores at depth and even producing a preliminary assessment of the content of copper in the primary ores.

The geochemical results determined that the Oyu Tolgoi porphyry systems were, at least locally, elevated in gold. A few rock samples from Central, Southern, and Western Oyu were auriferous, with gold content exceeding 1 ppm or 1 g/t gold (Fig. 3.27). Analyses of the rock samples from South Oyu yielded highly anomalous copper values (Fig. 3.28) because of the abundance of the copper mineral chrysocolla exposed on outcrops. Here the presence of gold at South Oyu was also notable. The average content of copper from 64 samples was 9,141 ppm or 0.91% and 0.28 ppm or 0.3 g/t gold. Results of geochemistry from West Oyu, in general, were comparable with the results of the Central Oyu. From 62 samples, the averages were 316 ppm copper and 0.06 ppm gold; however, unlike Central Oyu, the arsenic values were very low, averaging 6 ppm, compared to those of Central Oyu with 319 ppm arsenic (Fig. 3.29). Central Oyu also had the highest content of molybdenum (Fig. 3.30).

In general the difference between mineralized hills of Oyu Tolgoi in the geochemistry of nonferrous metals, such as lead and zinc, was the lowest in

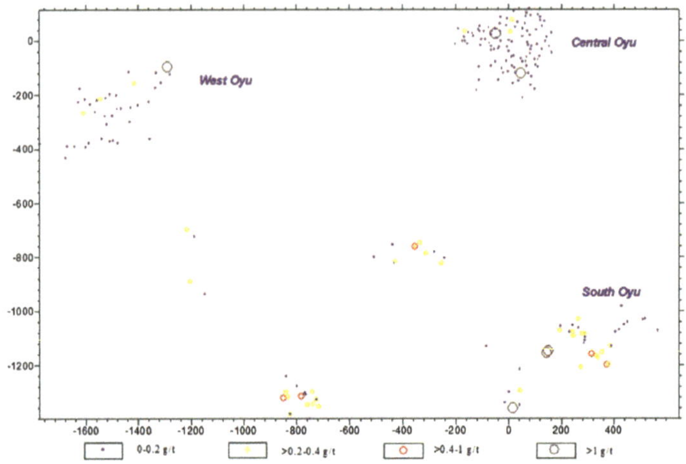

FIGURE 3.27 Distribution of gold in rock chip samples from Oyu Tolgoi.

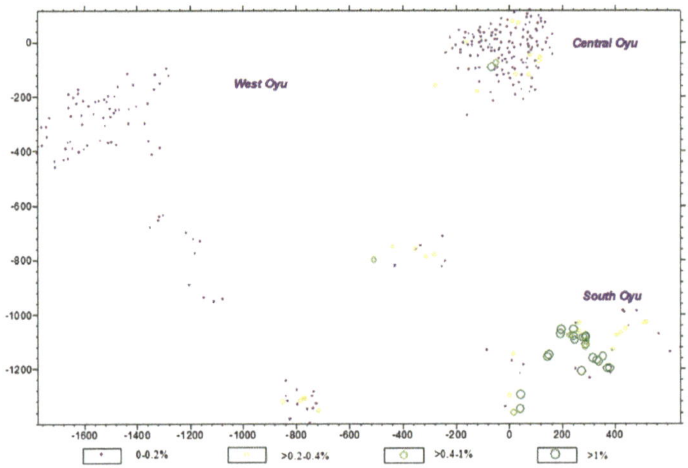

FIGURE 3.28 Distribution of copper in rock chip samples of Oyu Tolgoi.

Central Oyu (average of 19 and 22 ppm). The average content at South and West Oyu was significantly higher, 115 and 114 ppm for zinc and 62 and 30 ppm for lead. That is, generally speaking, there was particular geochemical zoning in mineralization seen in the outcrops of the Central, Southern, and Western Oyu Tolgoi hills. Built on the assay results, the maps of the distribution of copper, gold, molybdenum, and arsenic at Oyu Tolgoi hills revealed such patterns (Figs. 3.27–3.30).

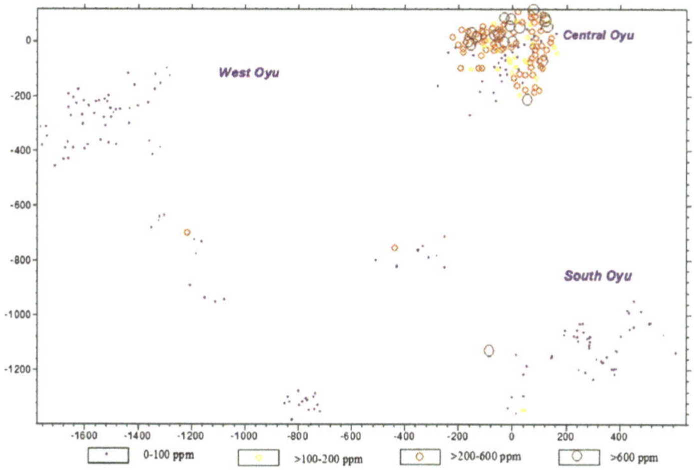

FIGURE 3.29 Distribution of arsenic in rock chip samples of Oyu Tolgoi.

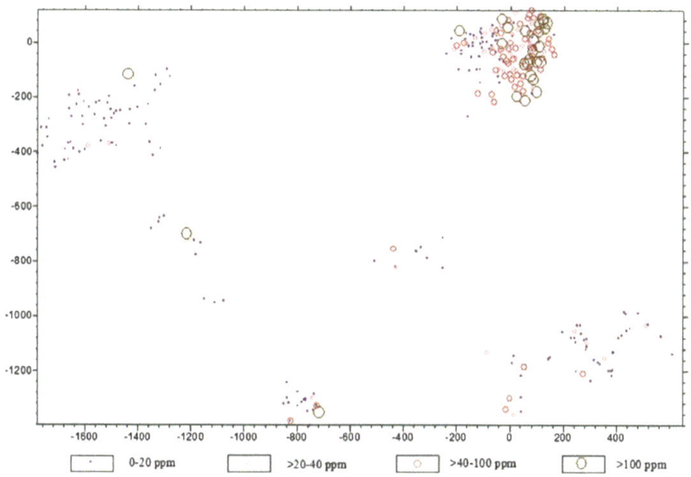

FIGURE 3.30 Distribution of molybdenum in rock chip samples of Oyu Tolgoi.

3.8 RESULTS OF COMPLETED SURVEYS

3.8.1 Central Oyu Tolgoi

On the western slope of Central Oyu, within a zone of well-developed advanced argillic alteration, a strong stockwork zone occurs and is described in a previous section. The orientation of quartz veinlets within the stockwork was random, with angles of dip ranging from subhorizontal to steeply vertical.

The study of limonites within the quartz veins showed that among iron oxides, goethite and jarosite were dominant. In the central part of the phyllic alteration zone, the goethite content ranged from 50% to 90%. Jarosite, ranging from 50% to 85%, prevailed within the zone of advanced argillic alteration. Limonite-rich jarosite occurred in the southeastern parts of the outcrops.

The quantitative ratio was apparently difficult to assess in those places where the hematite content was less than 10%. The sites with more than 20% hematitic were found both at the northwest flank of the outcrops and within the southern part of the phyllic alteration zone with quartz stockworks. Apparently the studies of limonites positively indicated a possible presence of the chalcocite blanket beneath Central Oyu. While conducting the property geological mapping the team also collected geochemical samples for analysis by Chemex lab in Vancouver (Fig. 3.31).

For the relic sulfide studies, 18 samples were collected from quartz veins. These were sent to Langerfeldt and Aguilar Ltd. in Santiago, Chile, for encapsulated sulfides studies. Minor remnants of sulfides, which were entirely encapsulated and preserved in quartz avoiding leaching, should be indicative of the hypogene sulfide mineral composition at depth. A. Aguilar found that half of the samples contained over 15 grains of sulfides per square centimeter. In the second half of the samples the number of sulfide inclusions reached more than 31. Pyrite was by far the dominant sulfide, ranging from 50% to 85% of the total sulfide grains, followed by chalcopyrite, chalcocite, bornite, and covellite in descending order (Figs. 3.32 and 3.33).

Interestingly, A. Aguilar concluded that, judging by their appearance, chalcocite was likely of primary origin because normally secondary chalcocite forms rims around grains of pyrite. Another persuasive argument is that if it

FIGURE 3.31 Mapping team at Oyu Tolgoi camp with the first batch of collected geochemical samples.

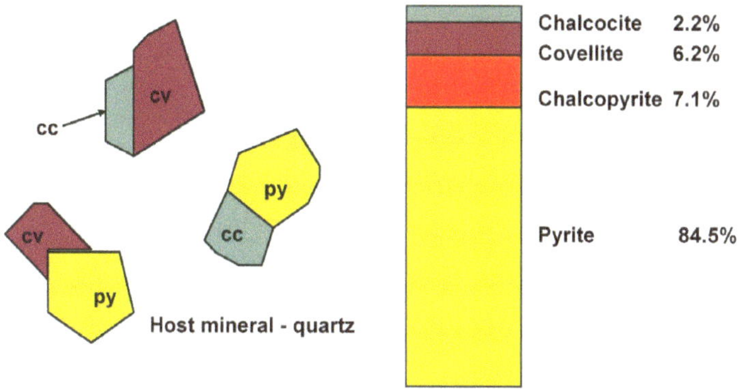

FIGURE 3.32 Results of relic sulfide studies in the sample B−193S. *cc*, chalcocite; *cv*, covellite; *cp*, chalcopyrite; *py*, pyrite.

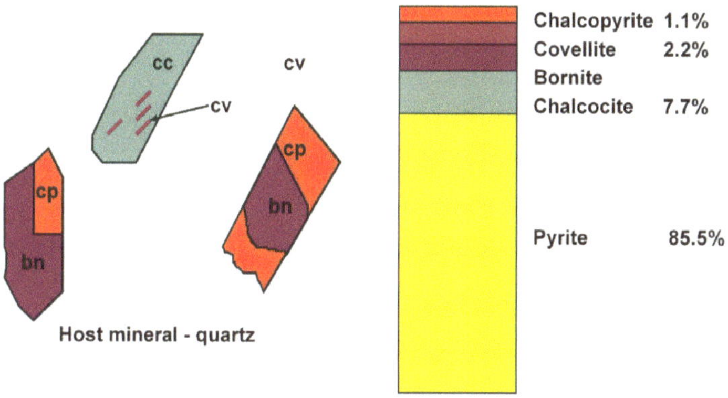

FIGURE 3.33 Results of relic sulfide studies in sample B−210S. *bn*, bornite; *cc*, chalcocite; *cv*, covellite; *cp*, chalcopyrite.

were secondary, then it would not be able to penetrate resistive capsule shells of quartz. The study of polished sections showed that in the ores of Central Oyu, there were at least two generations of chalcocite, one in the form of veinlets and the other as rims around pyrite. The first generation obviously had a primary origin, whereas the second generation was due to the process secondary sulfide enrichment.

There was no significant presence of arsenic minerals. The composition of pyrite did not show a noticeable amount of arsenic either despite the elevated content of arsenic in geochemical samples. In other words the mineralization promised to be free from the substantial presence of deleterious elements. Interestingly, the studies conducted under electron scanning microscope in the samples from the drill core did not show either presence of enargite or luzonite—common arsenic minerals in the upper parts of porphyry systems.

The investigation revealed only one grain of cassiterite and sulvanite. The latter had a similar formula to enargite, but instead of arsenic, it contained vanadium.

Geochemically, the Central Oyu Tolgoi Hill contained elevated concentrations of arsenic, copper, gold, and molybdenum. The copper content varied from 0.1% to 1.2%, with the highest content within dikes of andesites and the lowest contents in the zones of argillic alteration. The zone of intensive silicification and advanced argillic alteration yielded the highest gold contents, from 0.05 to 2.68 g/t, and also the most arsenic, from 200 to 1,700 ppm. Elevated molybdenum, in excess of 100 ppm, was a characteristic of the outcrops on the eastern slopes.

Correlation among the abovementioned elements reveal that the closest correlation, about 70%, exists between copper and molybdenum, whereas the correlation between copper and arsenic and between gold and arsenic were insignificant, at about 30%.

Geophysically, Central Oyu Tolgoi stood out above all other prospects primarily because of the sizeable magnetic low anomaly and overlapping high chargeability of IP anomaly (up to 13%).

D. Garamjav mapped South Oyu Tolgoi in April 1997. He had first visited the area in 1983 during his trip to Khanbogd Alkaline Complex. At South Oyu Tolgoi, mineralization was found on a small ridge located 1,100 m south of Central Oyu in two outcrops 700 m × 400 m and 400 m × 200 m in size. Each was composed of crystalline volcanic rocks with phenocrysts of feldspar.

Dikes of altered syenite porphyry with phenocrysts of potassium feldspar and hornblende intruded crystalline volcanic rocks. Syenite dikes appeared to be more altered on the western flank and to a smaller extent on the east. One of these dikes was cross-cut by a dike of unaltered rhyolite about 30 cm in width. Dikes of andesites also occur in the area. The potassic alteration was found in all types of rock units except the small rhyolite dike. The volcanic rocks contained fine-grained biotite and chlorite, probably hydrothermal products of alteration, replacing hornblende. Magnetite is also widely disseminated and may also be a by-product of hydrothermal alteration. Phyllic alteration was found only in the rhyolite dikes in the western part of the area. Quartz-magnetite veins and quartz veins, with hair-thin hematite in the center, formed stockworks with up to 50 veinlets per square meter. The stockwork veinlets were preferentially orientated in an east-west direction.

Within South Oyu outcrops, there was visible copper mineralization in the form of multiple films of malachite and chrysocolla along cracks. Turquoise was seen only at one location.

3.8.2 North Oyu Tolgoi

We first mapped the outcrops of North Oyu in April, 1997. It is a small hill about 100 m in diameter about 600 m north of Central Oyu. It is a part of the same magnetic destruction anomaly as Central Oyu, and lies within the northern tip of the IP anomaly. The hill consists of volcanic porphyry rocks that were hydrothermally altered and intensely silicified, accompanied with associated alunite, dickite, and zunyite. Leached pockets and cavities were noted in the silicified zones. There was no clear indication of quartz veining. Iron oxides were represented by limonite, which was slightly more common than goethite and jarosite combined (3:2 ratio).

Six rock samples returned copper values from 59 to 269 ppm, arsenic from 16 of 284 ppm, and molybdenum from 12 to 56 ppm. Gold and silver in the samples were detected in small quantities of 0.02 and 0.4 ppm, respectively.

No noteworthy geophysical anomalies emerged from the surveys at North Oyu.

3.8.3 Mineralization at South and South-West Oyu Tolgoi

South-West Oyu is located to the South-West of Central Oyu, covering an area of 1,000 m × 600 m. It is the western extension of the South Oyu magnetic anomaly. Mapping at 1:25,000 scale conducted in June of 1997 delineated southwest. The area was found to be composed of altered volcanic rocks intersected by numerous syenite and syenite-porphyry dikes and to a lesser extent by aplite dikes. In the south, dikes of rhyolite and andesites along with aplite also occur.

Here hydrothermal alteration is dominated by potassic metasomatism resulting in alteration of hornblende to biotite in the syenite-porphyry dikes. The rocks in the south flank of this prospect are rich in magnetite.

Mineralization at South-West Oyu is in the form of primary chalcopyrite, which is found predominantly in quartz veinlets trending either in northeastern or northwestern directions. The veinlet density is considered to be moderate and ranges from four to seven per square meter. In the northern flank of South-West Oyu Tolgoi, there were several outcrops of quartz veining with patches of secondary malachite. These outcrops were marked by stone piles (Fig. 3.7). Connecting them by imaginary lines on the map would make the shape of a triangle with the sides a few hundred meters long. The veinlets had a clear middle line with magnetite that could contain higher gold content. This imaginary triangle became to be known as the "golden triangle".

South-West Oyu Tolgoi is anomalous both in copper and gold. The copper content was relatively low, ranging from 0.1% to 0.3%, and anomalous gold values are between 0.1 and 0.97 g/t. The correlation between copper and gold is strong at 63%. IP was with a chargeability of 3%.

3.8.4 Additional Prospects within Oyu Tolgoi License Area

Based on the results of the reconnaissance work within the limits of the Oyu Tolgoi license area, geologists recommended a set of follow-up work at several newly identified mineralized prospects. New copper prospect was identified at a distance of 8 km south–southeast from Oyu Tolgoi on the southern bank of dry wash Undain Gol. The main credit for discovery of this prospect belonged to T. Munkhbat. In June 1997, when passing through the area, he noticed outcrops of limonite manifestations. The name of the prospects with blue rock (Fig. 3.34) was Khokh Khad (blue rock in Mongolian).

The area of Khokh Khad prospect was composed of gray andesite lavas, tuff-breccias, and dikes. One of the banded rhyolite dikes just forms the blue rocks on the northern flank of the ore, which gave rise to its name. In the southeastern flank, outcrop intrusives of syenite-porphyry were mapped. In the south, andesite volcanics were protruded by granites, which were not affected by hydrothermal alterations. Two areas of the mineral prospect were affected by hydrothermal alterations. The northwest flank showed developed zone of argillic alteration with dickite and pyrophyllite. A dike of unaltered and andesites intersected this zone of alteration. In the southwestern flank during mapping the geologists outlined a zone of argillization with silicification and argillization with dickite. Outcrops of syenite were altered to illite.

Quartz veins were quite rare, but on the northeast flank, there was a well-developed stockwork with veinlets of quartz and limonite, counting up to 25 veinlets per square meter (Figs. 3.34 and 3.35). Limonite composition dominated by the presence of jarosite. Hematite was also present in the amount of 10% of the rock volume. On the surface, there was no obvious presence of copper minerals. A tectonic fault was extending along the Undain Gol apparently with some offsetting. Volcanic rocks with syenites on the opposite bank were not affected by hydrothermal alterations, which confirmed the idea of dislocations along the alleged fault.

Shivee Ovoo was another mineral occurrence identified within the area of Oyu Tolgoi. In July 1997 during mapping with Portable Infrared Spectrometer (PIMA), Dr. Cox identified the presence of advanced argillic alteration here.

FIGURE 3.34 Prospect Khokh Khad with a view on herdsmen's gher and cattle corral.

FIGURE 3.35 Weakly pronounced stockwork at Khokh Khad occurrence.

From 50 samples collected within the mineral occurrence, 23 showed presence of gold (more than 0.05 ppm), with 5 samples exceeding gold content 0.1 g/t, and one of the samples taken from the sericitized volcanics yielded elevated gold content (0.98 g/t) and silver (2 g/t), as well as anomalous presence of arsenic, copper, lead, and zinc.

Mineralization was located at a distance of 11 km north–northwest from Oyu Tolgoi. Area of altered rocks stretches for a distance of 2,500 m and occupies approximately 1,000 m in width. It may have received its name due to presence of an ancient observation post (Shivee) on top of a hill (Ovoo) in the center of the zone of hydrothermally altered rocks (Fig. 3.36).

The host rocks were represented by rhyolites and andesites specific for eastern flank area, whereas on the western flank dominated granite porphyries, gradually turning into medium-grained granites in a westerly direction.

Hydrothermal alteration was caused by substitution of volcanic rock with silica material (Fig. 3.37). Silicified rocks had numerous voids and indicated presence of brecciation. Based on PIMA measurements (Fig. 3.38), in the

FIGURE 3.36 Panorama of Shivee Ovoo occurrence. The outcrops are composed of granites (left) and andesites (right). The rocks in the middle are silicified volcanics with alunite.

FIGURE 3.37 Quartz-alunite rock with jarosite found in outcrops of Shivee Ovoo occurrence.

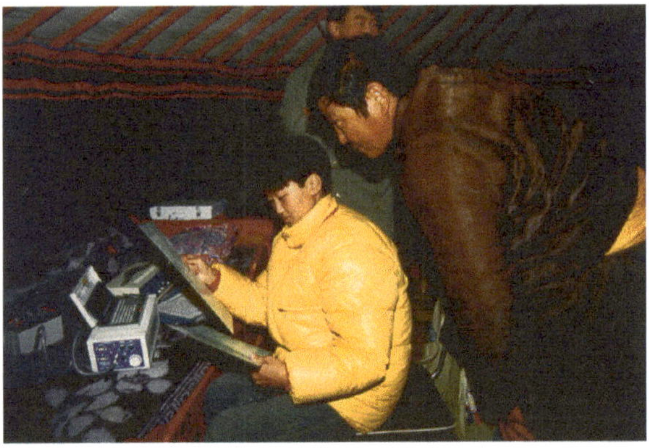

FIGURE 3.38 Oyun and Sanjdorj discuss the results of PIMA measurements.

outcrops, besides alunite, zunyite and topaz were also detected. The dominant clay mineral in hydrothermally altered andesites was illite.

Silicification zone was shown as lenses rather than quartz veins. At 150 m to the southeast from the top of the Shivee Ovoo prospect, an outcrop with stockwork was found, where the density of quartz veinlets ranged from four to five veinlets per square meter. In altered rocks, especially within the silicified intervals, numerous patches of limonite were seen. Regarding constituent minerals of limonite, goethite, and jarosite, they were present in approximately equal quantities with only traces of hematite. The presence of distinct copper minerals on the surface was not established. From the 39 geochemical samples taken here, two samples returned elevated gold content reaching 0.045 and 0.19 ppm with copper and arsenic contents at very low levels, respectively, 109 and 154 ppm.

Both Khokh Khad and Shivee Ovoo on the surface manifested indications of possible presence of copper porphyry mineralization at depth. The absence of intensely developed quartz stockwork indicated that the copper mineralization here was apparently at a greater depth than that at Central Oyu. To verify the presence of sulfide mineralization, it was suggested to conduct IP geophysical survey. If the depth of geophysical anomaly was within the reasonable range for open pit mining, then it was suggested to subsequently test these anomalous sites with diamond drilling.

Chapter 4

Drilling and Resource Calculation Results

Chapter Outline

Barrie Bolton*—graduate of the Monash University in Melbourne, Australia, had PhD from the same university. Earlier in his career, he worked on manganese deposits, also supervised exploration program for gold in Ghana, participated in exploration of copper–gold deposit Ok Tedi in Papua New Guinea. At the time of his appointment, he was responsible for the safety of Eurasia geological unit. From 1998 to 1999, he led the exploration of the BHP program in Mongolia. Currently, he is a honorary researcher at Monash University, Melbourne, Australia.*

4.1 FIRST STAGE DRILL PROGRAM PLANNING

In August 1997, BHP Minerals contracted a local company Gobi Geo to conduct drilling on the Oyu Tolgoi license. The planned program was small consisting of 1,000 linear meters of diamond core drilling. All participants provided their suggestions for drill target sites. A total of six anomalous sites

Discovery of Oyu Tolgoi. https://doi.org/10.1016/B978-0-12-816089-3.00004-4

were on the list for drill testing: Central Oyu, South Oyu, South-West Oyu, North and North-West Oyu.

Immediately after signing the contract agreement, the Gobi Geo drilling team moved into the field with two ABS drilling machines ZIF-650 and SBA-500. Both of these Soviet-made machines had the capability of diamond core drilling to a depth of 500 m. Explorers in Mongolia had successfully used these machines for exploration drilling of copper, gold, and fluorite deposits in the past. The SBA-500 ABS machine could drill only vertical drill holes, whereas the ZIF-500 machine could drill both vertical and inclined drill holes, although the angle was limited to 75°.

This exploratory drilling phase was designed to test the highest priority targets, which had been identified based on the data obtained from the geological mapping, exploration geochemistry, and geophysical surveys (Fig. 4.1, Table 4.1).

Drill hole OT01 was targeted at testing the most intensive IP anomaly on the project with a maximum chargeability of 13%. Drill hole OT02 was designed to test another IP anomaly under cover with a chargeability of 5% in northwestern flank of Oyu Tolgoi property. The objective of drill hole OT03 was to test for the presence of a zone of secondary enrichment underneath Central Oyu.

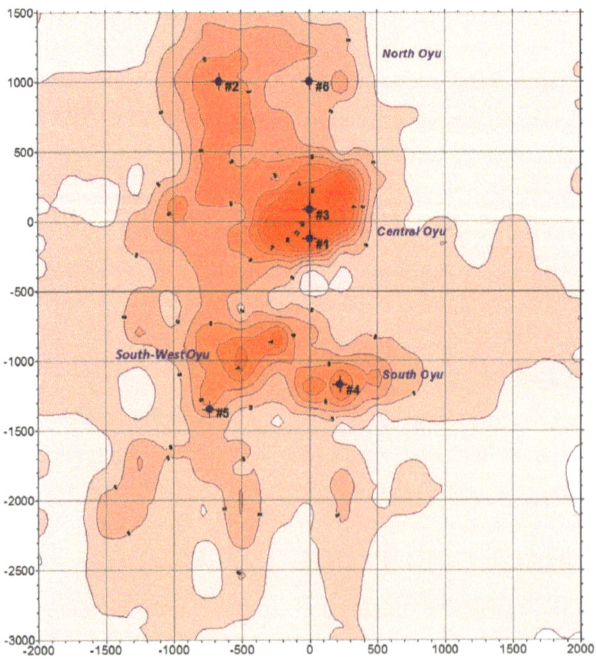

FIGURE 4.1 Location of drill holes completed during the first drilling stage of 1997 field season (IP results with chargeability anomalies in the background).

TABLE 4.1 List of Drill Holes Completed at Oyu Tolgoi During the First Stage of Drilling in 1997

	Drill Holes					
	OT-01	**OT-02**	**OT-03**	**OT-04**	**OT-05**	**OT-06**
Northing	43°00′49.6″	43°01′25.3″	43°00′56.4″	43°00′16.7″	43°00′13.1″	43°01′24.6″
Easting	106°50′56.4″	106°50′26.4″	106°50′58.0″	106°51′04.5″	106°50′26.1″	106°50′57.6″
Elevation, m	1,166	1,166	1,168	1,168	1,166	1,160
Direction	350°			40°	226°	
Angle	75°	90°	90°	75°	75°	90°
Depth, m	136	98.9	186.3	250.7	207.9	121.4

When considering possible drill targets at both South and South-West Oyu Tolgoi, a concern was that, although there was exposed copper, the grade seemed to be low and the evidence for leaching, mostly in the form of weak oxidation, was minimal. Hence, there is virtually no chance for secondary enrichment at depth. A consensus was formed that neither South nor South-West Oyu deserved further attention due to a simple reason that based on the principles of "encapsulated sulfide studies," the tenor of copper in the hypogene zone at depth should be similar to the content of copper on the surface. The catchphrase "What you see is what you get" accurately reflected those concerns.

S. Diakov, the project leader, began to have reservations about this decision after reflecting on the Grasberg gold–copper deposit and the fact that South-West Oyu Tolgoi contains significant gold values. Grasberg is among the richest mines in the world, which Sergei had visited the year before. He was intrigued by what a profound impact an added "sweetener" like gold could have on the value of a copper porphyry style deposit, even without the benefit of a supergene zone. Despite some objections from the team among members who wished to remain faithful to the supergene model, two drill holes, OT04 and OT05, were ultimately designated to explore for primary copper–gold mineralization at depth. The location of the last drill hole on the list, OT06, was at North Oyu to test the weak IP anomaly.

4.2 FIRST STAGE DRILL PROGRAM EXECUTION

D. Garamjav and S. Sanjdorj supervised the drilling program. They were responsible for producing a geological log of the drill core producing core logs at a scale of 1:100 and also participated in the geochemical sampling of Oyu Tolgoi and the adjacent sites. Drilling continued until the first frost by which time six holes totaling 1001.2 linear meters had been completed. Table 4.1 summarizes the list of drill holes at Oyu Tolgoi completed during the first stage of drilling.

On completion of a quick core log and the recording of the technical core measurements (Fig. 4.2), the drill core was transported to Ulaanbaatar. There the drill core was cut into half lengthwise by a diamond saw. One-half of the core was used for assay sampling from each 2-m interval and was shipped to the laboratory and the other half was "core-logged" (i.e., the core was thoroughly studied and the data were documented by a geologist). The geological logging procedures that were developed during the Escondida exploration-drilling program were applied at Oyu Tolgoi with a description of the core every 2-m interval.

As drilling progressed, in September, Senior Vice President of BHP Minerals Hugo Dummett, Regional Manager of BHP North Eurasia Donald Schissel, safety coordinator of North Eurasia Region Barrie Bolton, and the Branch Manager of BHP Minerals in Mongolia Sergei Diakov visited the BHP

FIGURE 4.2 The first core boxes with drill core from Oyu Tolgoi. Garamjav and Sanjdorj are conducting geological logging of the core.

field camp. By this time, two drill holes OT03 and OT04 were completed. The timing of this field visit by management could not have been better because the drill core boxes laid out on the campground and in process of being logged by the geologists, which happened to be the discovery drill holes at Oyu Tolgoi.

In drill hole OT03 from Central Oyu Tolgoi, secondary chalcocite enrichment was intersected, and in drill hole OT04 from South Oyu, the mineralization consisted of primary chalcopyrite–bornite mineral assemblage. This confirmed that at Oyu Tolgoi, there is a mineralized system with a variety of types of copper mineralization from secondary copper enrichment to primary copper–gold sulfide mineralization. Drill hole OT03 intersected 10.1 m of 2.39% copper in a supergene blanket and another 16.9 m interval of 1.89% copper. Drill hole OT04 intersected a 70-m interval grading 1.65% copper and 0.28 g/t gold.

These intercepts were encouraging for the program and were in line with our expectations for discovery of secondary copper enrichment at Oyu Tolgoi. At this early stage in the program, the discovery of a huge chalcocite blanket was in our primary focus. Seeing these high-grade copper drill intersections first hand at such an early stage of the project was a confidence booster for the exploration team and made a favorable impression on management.

BHP's exploration work in the Gobi Desert attracted the attention of the Mongolian authorities. In September 1997, a delegation from the Mongolia Cadaster Office led by D. Jargalsaikhan (Figs. 4.3, 4.4 and 4.5) visited the Oyu Tolgoi site.

BHP's program confirmed that the adopted laws governing foreign investments were also effective for the mineral sector and that foreign companies were beginning to invest actively in geological exploration of Mongolia. The discovery of Oyu Tolgoi was possibly one of the first rewards for opening up Mongolia to foreign investors.

FIGURE 4.3 The first drilling rigs at Oyu Tolgoi: SBA-500 on the left and ZIF-650 on the right.

FIGURE 4.4 Visit of Oyu Tolgoi camp by the delegation of the Cadaster Office during drilling in September 1997.

4.3 DISCUSSION OF FIRST STAGE DRILL PROGRAM RESULTS

Below is a brief description of the drilling results included here for the geologically inquisitive readers. Borehole OT01 in Central Oyu was drilled to a depth of 136 m. At 27-m downhole, the drill intersected a fault zone and

FIGURE 4.5 Review of drill core. *From left to right above: S. Diakov, D Jargalsaikhan, and drill supervisor of Gobi Geo; below left to right: S. Sanjdorj, D. Garamjav, and G. Oyun.*

remained in the structure until the 86-m mark. Drilling conditions in this interval were challenging because the rocks were severely fractured. The rocks in this hole were intensely altered containing fine-grained phenocrysts of quartz and feldspar. Quartz—sericite alteration assemblages were ubiquitous. At the surface level, aggregates of dickite occurred, and in the cap rock, kaolin was the dominant supergene alteration product but with depth pyrophyllite started to appear in the drill core. Mineralization in the hole was in the form of scattered and dispersed D-type veinlets of pyrite in association with chalcocite and covellite (see Annex 4 for vein-type explanations). In the 16.7—52 m interval, the average copper grades were 0.8%. A-type veins occur in the central part of the interval with pyrite veinlets. The younger type D veins are crosscutting. They are typically 1 cm thick and contain pyrite with quartz margins along veinlet boundaries. These veins apparently carried the main volume of the secondary mineralization. The total number of sulfides varies between 3% and 10%, averaging 5% of the volume of rock. At 60.1 m downhole, drilling intersected the postmineral rhyolite dike, which was less than a meter thick.

Borehole OT02 was drilled to a depth of 98.9 m in North-West Oyu. Initially, it passed through the overlying cover rock, most likely Quaternary in age, consisting of red clay on top and then interbedded clays followed by gravelites down to 54 m. The host rocks below this level were composed of pinkish-green porphyritic syenodiorites with phenocrysts of potassic feldspar and plagioclase ranging in size from 3 to 4 mm. Mafic minerals, predominantly hornblende replaced by chlorite, constitute 10% of the volume. The matrix was fine-grained indicating that emplacement was subvolcanic. In the 41.3- to 46.5-m interval, the drilling penetrated hydrothermal breccia, consisting of abundant quartz fragments with a weakly crystallized cement of chlorite and albite. In the zone of brecciation, crystals of biotite were present.

The alteration is dominated by intense silicification, chloritization, and albitization. The drill hole intersected a mineralized interval of 14 m with an average content of 1.09% copper and 0.23 g/t gold, coinciding with the zone of brecciation from 66 to 80 m. Disseminated inclusion of chalcopyrite with traces of bornite in albite adjacent to quartz veinlets represented the dominant type of mineralization.

The OT03 hole was drilled at Central Oyu. Initially, the drill hole went through 2 m of cover rocks and then entered into the brecciated feldspathic rocks. The feldspathic rock consisted of 20%–30% anhedral to subhedral feldspar phenocrysts ranging in size from 1 to 2 mm submerged in a matrix of quartz and sericite. Mafic minerals compose less than 5% of the rock and have been completely replaced with secondary minerals. Fragments of feldspar and quartz, from a few to more than 10 cm in size, represent hydrothermal breccia. The cement revealed some fluid streak textures, consisting of grains of quartz, feldspar, and sericite–dickite with disseminated impregnations of sulfides. Randomly oriented stockwork quartz veins often cut fragments of feldspar. The chalcocite mineralization was intersected at a depth of about 20 m (Fig. 4.6). In this drill hole, core recovery was quite poor due to the outdated drilling equipment we used at that time.

The drilling penetrated the upper part of the silicified leach zone, which contained a considerable amount of hematite. This conformed precisely to the secondary enrichment model. Below this zone, the drill hole intersected a light blue–colored zone with abundant secondary chalcocite mineralization, which enveloped pyrite with numerous coatings. In the 63- to 70-m interval, drilling penetrated a brick red–colored, postmineral, medium- to coarse-grained porphyry dacite dike, which apparently cut both hydrothermal breccias and mineralization. The amount of chalcocite in OT03 exceeded 10% in places (Fig. 4.7), which demonstrated that there was good potential for a large volume of secondary sulfide enrichment of ore grade.

FIGURE 4.6 Core of OT03 drill hole. Red core indicates mineralization with hematite representing lower part of the leached zone. The light blue core represents chalcocite mineralization in the zone of secondary enrichment.

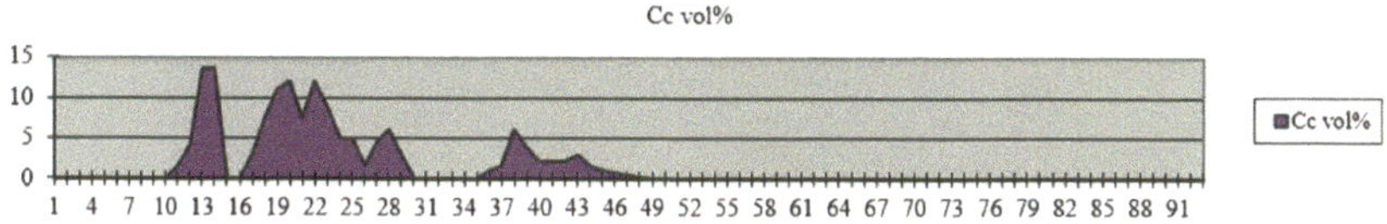

FIGURE 4.7 Distribution of chalcocite (cc) in drill hole OT03. The horizontal axis marks 2-m intervals and the vertical axis indicates the volume percent of chalcocite.

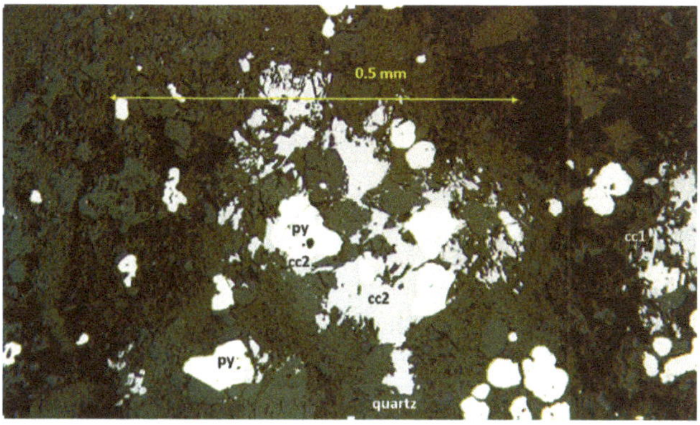

FIGURE 4.8 Pyrite (py)—chalcocite (cc) mineral association. Two generations of chalcocite are present. Fissures in pyrite carry chalcocite 1 (tentatively digenite). Chalcocite 2 replacing pyrite. Sample from OT03 drill hole at 42-m interval.

In polished sections, it was apparent (see Fig. 4.8) that chalcocite formed chalcocite rims around pyrite, a typical symptom of secondary sulfide enrichment. At a depth of about 150 m, the mineralization contained covellite, which formed separate aggregates and apparently had a primary petrogenesis. In the upper part of the core, the dominant type of alteration was supergene and hypogene alunite. Supergene alunite usually occurs at the top of the leached zone of the porphyry system, forming millimeter to centimeter thick veinlets. Hypogene alunite is white to pink in color and appears to replace feldspar and fill the voids in brecciated rock. Along the cracks, white dickite developed throughout the core, which at the bottom of the drill hole is associated with pyrophyllite. The matrix of the rocks consisted of quartz and sericite. Hypogene kaolinite was developing along the tectonic faults. Most of the mineralization in the borehole occurred in the chalcocite blanket, where steely colored chalcocite developed by replacing disseminated sulfides, primarily pyrite.

At the top of the chalcocite blanket, the richest copper zone, grading up to 2.39% copper coincided with massive sulfide D-veins in the interval from 20.7 to 30.8 m at the top of the chalcocite blanket. Below, in the interval from 30.8 to 50 m, a leached zone with hematite and pyrite contained 1.89% copper. Starting at 54 m downhole, covellite first appeared and at depth became the

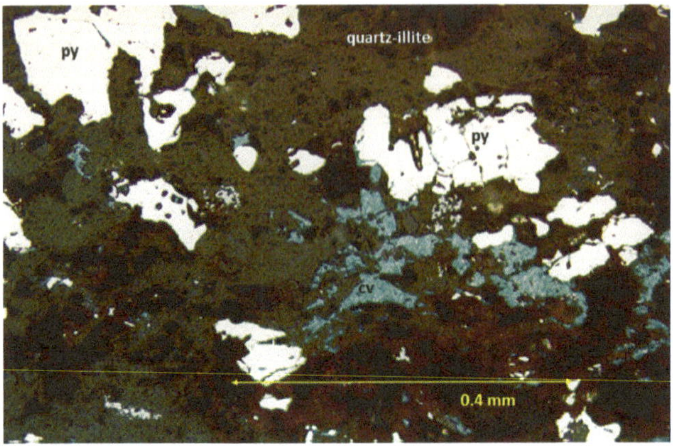

FIGURE 4.9 Pyrite (py)—covellite (cv) mineral association. Sample from drill hole OT03 at 153.8 m.

dominant copper mineral, especially below 90 m (Fig. 4.9). The type A and type B veinlets were present, but they were almost devoid of sulfides. D-veins carried hypogene chalcocite and covellite with associated alunite with dickite. The total number of sulfides ranged from 2% to 7% averaging 4.5%.

It was overly exciting to see the model working effectively. The team was very enthusiastic and continued working long hours to complete the drilling program before the cold season gripped the vast steps of the Gobi.

Borehole OT04, in South Oyu, was drilled to a depth of 250.7 m. The drill core for the most part was intensely fractured, especially below 44 m, due to the presence of tectonic faults (Figs. 4.10 and 4.11). The host rock, andesite porphyry contained numerous dikes of andesite and syenite. The andesite was

FIGURE 4.10 Core of drill hole OT04 transitional zone of oxidation to reduction with native copper.

FIGURE 4.11 Core of drill hole OT04 with bornite mineralization.

typically dark green containing about 20% feldspar phenocrysts and about 5% of mafic phenocrysts in an aphanitic matrix. In places, the andesite is marked by striations with a perpendicular orientation relative to the axis of the core. The syenite dikes are essentially unaltered and contain about 5% phenocrysts. Throughout the drill hole, the selvages of the quartz–sulfide and magnetite–sulfide veinlets are altered intensely to chlorite and a lesser volume of albite and relics of biotite. The albite has the appearance of whitish, amorphous material.

The typical mineralized assemblages in borehole OT04 are quartz–chalcopyrite–magnetite, magnetite–chalcopyrite–bornite, bornite–magnetite, chalcopyrite–magnetite, and chalcopyrite–bornite as veinlets. In addition, chalcopyrite also occurs as disseminations along albite-chlorite salvages around the quartz–sulfide veins. In the magnetite and bornite-bearing intervals, the rock was unusual dark, almost black, with tints of blue (Fig. 4.11).

Multidirectional quartz–magnetite–sulfide veins forming stockworks dominate the strongly mineralized interval between 56 and 126 m with 1.65% copper and somewhat elevated gold (0.154 g/t). Some of the veins paralleled the axis of the core. In this interval, the total sulfide content varied from 1% to 7%. In the zone of oxidation above the depth of 52 m, the copper content was low averaging 0.28%. Tenorite and malachite were dominant copper minerals. The interval from 52 to 56 m contained 0.95% copper and 0.57 g/t gold mainly in stringers and veinlets of native copper predominantly oriented at high angle to the core (Fig. 4.12). These represented a zone of cementation.

Below the zone of cementation, the primary mineralization was represented by aggregates of bornite with chalcopyrite in a quartz–biotite matrix with inclusions of hematite. In the interval between 56 and 126 m, bornite was the dominant copper sulfide mineral (see Fig. 4.13) contributing to relatively high copper and gold grades (Fig. 4.14). Near the bottom of the hole at 214 m,

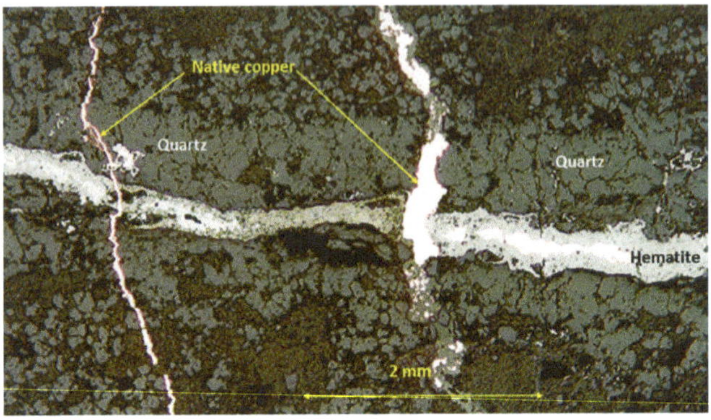

FIGURE 4.12 Polished section with veinlets of native copper crosscutting quartz–hematite vein. Drill hole OT04, 54.6 m.

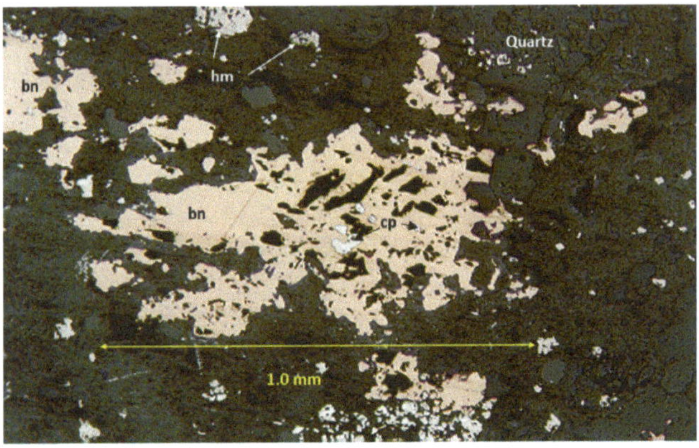

FIGURE 4.13 Aggregates of bornite (bn) with chalcopyrite (cp) in quartz–biotite cement with inclusions of hematite (hm). Drill hole OT04, 63.7 m.

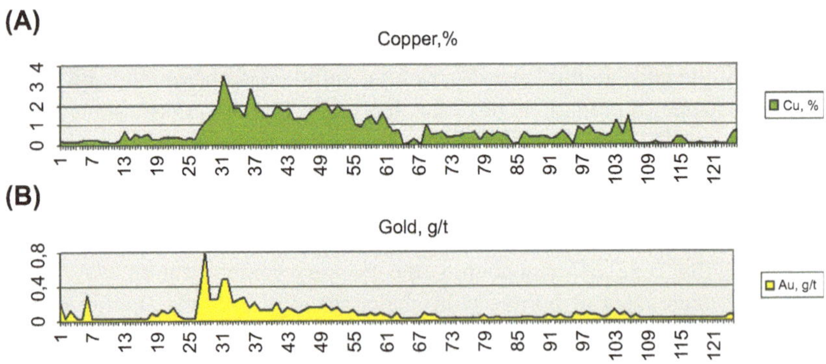

FIGURE 4.14 Distribution of copper (A) and gold grades (B) along drill hole OT04. Horizontal axis indicates 2-m intervals.

pyrite became more dominant, where it was closely associated with minor chalcopyrite resulting in very low copper and gold grades.

Borehole OT05, at South-West Oyu, penetrated the dark-colored andesites below the Quaternary cover 13 m downhole. These andesites are a fine to medium grained, weakly porphyritic crystalline rock containing 5%–10% pyroxene and hornblende phenocrysts. Some fluidal texture was preserved. Dikes of andesite porphyry with a fine-grained matrix containing cavities and voids filled with calcite were similar to the andesite dike described in borehole OT04. Apparently, these dikes formed in postmineral time since they were only weakly pyritic. The prevailing hydrothermal alteration was chloritization with locally preserved small flakes of biotite. Pyrite and chalcopyrite occurred as patchy quartz–pyrite–molybdenite veins tended to be haloed by chalcopyrite-rich disseminations, whereas pink albite surrounded veins with molybdenite. Sometimes the quartz–pyrite–chalcopyrite–molybdenite veins contained small inclusions of fluorite. The total sulfide content ranged from 1% up to 6% and averaged about 2%. In places, the veins were oriented along the core axis. The best-mineralized interval in this hole is between 160 and 184 m averaging 0.96% copper and 1.14 g/t gold.

Drill hole OT06, just west of North Oyu, was 121.4 m deep. At the top, it intersected a thin cover of Quaternary gravelites, which persisted to 26.2 m downhole. Below the cover, the drill hole transected thinly banded sandstones and siltstones. Below the 80-m mark, the sedimentary rocks contain fragments of syenites. In places, there were traces of pyrite and weak indications of chloritization. Apparently, the drill hole OT06 did not reach the volcanic rocks and was stopped in the sedimentary cover.

4.4 SUMMARY AND CONCLUSIONS OF FIRST STAGE DRILL PROGRAM

The drilling results and the first mineral intercepts were the first steps toward understanding the mineralized porphyry system at Oyu Tolgoi. Below is a summary of the preliminary conclusions drawn from the results of the first stage of drilling.

The best mineralization was intersected at Central Oyu Tolgoi by holes OT01 and OT03, which tested a coincident IP anomaly with a chargeability strength of up to 13%, the strongest IP anomaly on the Oyu Tolgoi license. This was a zone of secondary sulfide enrichment, 24–30 m thick, containing chalcocite and to lesser extent covellite. The zone of secondary enrichment is located over the center of a multiphase intrusion with breccias, which generated the copper-rich hydrothermal solutions. The supergene zone was intersected as close as 18 m below the surface. At deeper levels, the presence of aggregates of primary chalcocite and covellite associated with hypogene alunite indicated that these copper-enriched minerals had a primary paragenesis. Apart from introducing the copper, the hydrothermal solutions were also responsible for intense advanced argillic alteration. This alteration is interpreted to be a steeply plunging isometric tabular body which cross cuts the potassic alteration zone. The central part of the alteration zone contained predominantly alunite surrounded by zone of a dickite–pyrophyllite assemblage with an outer phyllic shell of quartz–sericite. The most intense alteration was in the zone with primary alunite. Here D-veins and A-veins were composed mostly of pyrite. These did not carry much copper, except in the zone of secondary sulfide enrichment, where supergene minerals precipitated producing a natural zone of supergene enrichment. Copper mineralization at depth appeared to have been developed over several generations. Later cupriferous solutions enriched the chalcopyrite to covellite and chalcocite. Central Oyu was considered a telescoped part of the porphyry system. It is believed that at an early stage of potassium-silicate metasomatic replacement, A-veins were formed, and during a later stage hydrothermal alteration quartz–pyrite, D-veins were generated in parts of the system. A-veins, at later stages, were superimposed with portions of the hydrothermal solutions, leading to the deposition of D-veins with quartz–pyrite. Subsequent hydrothermal solutions overprinted these veins resulting in the formation of advanced argillic alteration. The process culminated with the introduction of a considerable volume of postmineral rhyolite dikes and dacite dikes. These units were not associated with significant alteration and were essentially devoid of copper mineralization. At first, there was a serious concern that these barren bodies could create a significant dilution factor. At Central Oyu Tolgoi, surface rock chip sampling returned elevated arsenic suggesting that there is potential for enargite veining at depths probably not tested during phase 1 drilling. On the one hand, this would increase the chance for elevated copper content in the chalcocite blanket; however, this could potentially create some metallurgical problems. The drilling results confirmed that these fears were groundless; our detailed studies did not confirm significant presence of arsenic at depth.

As mentioned above, drill hole OT04, which tested the potential for a "Grasberg-style" zone of gold enrichment within the copper porphyry system at South Oyu Tolgoi, proved to be a pleasant surprise. The hole, which intersected an interval of 70 m grading 1.65% copper and 0.154 g/t gold, confirmed not only that there were significant gold credits at depth but also that the copper tenor was higher than expected and there was a strong correlation of gold mineralization with copper. It was suggested that this mineralization resembles the rich vein deposits of Butte in Montana or Lepanto in Philippines. The trend of the mineralized zone appeared to follow the chain of these small pits, described earlier. At South Oyu Tolgoi, the host andesitic volcanic rocks were intensely altered to an assemblage of chlorite–biotite–albite, believed to form due to potassic metasomatism. Chlorite appeared to be the result of retrograde metamorphism. Locally, silicification was developed in a quartz–magnetite stockwork with weak pyrite. Albite developed around magnetite—sulfide minerals and quartz—sulfide veinlets. The majority of chalcopyrite—bornite mineralization was associated with biotite, introduced by potassic metasomatism. Quartz—chalcopyrite mineralization was found in the peripheral and locally crosscutting retrograde phyllic quartz—sericite alteration zone. Despite the fact that the exposed syenite intrusions manifested signs of metasomatic potassic alteration, borehole OT04 did not intercept the type of porphyry intrusion that was drilled in Central Oyu. However, the presence of the high-temperature mineral assemblage of quart—magnetite—chalcopyrite—bornite, according to José Perelló, indicated a proximity of mineralization to the porphyry intrusion, which, of course, should be the source of mineralization.

Borehole OT05 confirmed that mineralization continued under the cover of postmineral rocks due to presence of chlorite—biotite and epidote—calcite here, which were interpreted as propylitic alteration. According to the classic copper porphyry model, this was the outer zone of the system. The elevated presence of molybdenum also pointed to a peripheral position in the system. The mineralization intercepted here partly resembled the style of mineralization intercepted by OT04.

The drill hole OT02 in the north presented to some extent an intriguing case. It was drilled on the margins of the outcropping prospects within the northern tip of the IP anomaly. The drill hole intersected a 14-m interval of chalcopyrite mineralization associated with alteration assemblages not found in the center of a porphyry system. This could infer that, similar to borehole OT-05, borehole OT-02 was situated in the outer part of the porphyry system.

Structural position of Oyu Tolgoi mineralization was underpinned by the fact that Central Oyu Tolgoi was located at the intersection of NW and NE striking tectonic faults, whereas mineralization at South Oyu by virtue of its position was controlled by a fault zone striking in northwestern direction. A series of tectonic faults cross cut and offset granite batholith intrusions as shown on the magnetic maps, as well as on the satellite images of Oyu Tolgoi.

The origin of these tectonic faults was most likely the results the tectonic block movement resulting in development of depressions and uplifts, extending along the arc, curved to the south from west to east along Mongolia. The block tectonic structure of the volcanic belt largely predetermined block tectonic structure of the Oyu Tolgoi area as well. The movements along the faults occurred in premineral and postmineral time. The long history of tectonic movements along these blocks consequently resulted in different elevation and position of the mineralization. Mineralization of South Oyu obviously of higher temperature formation was neighboring with the epithermal mineralization of Central Oyu, which again could be explained by significant block movement in the area during the postmineral time. We should note that the limited exposure due to development of postmineral cover significantly complicated our adequate interpretation of the mapping and the drilling results.

The magnetic results indicated that the block with reduced magnetization was confined by the mapped northwest and northerly trending faults. This block together with the mineralization was apparently down-dropped, which prevented the porphyry mineralization from being completely eroded.

To get a crude indication of whether phase 1 drilling results could imply that BHP's minimum ore tonnage requirements could be met even using best case scenarios, the geologists came up with some innovative ideas. In the case of Central Oyu Tolgoi, we speculated that the IP chargeability percentage could be used as a gauge to estimate the percentage of copper in the underlying rocks. If one accepts the assumption that the highest copper content of 1.5% or higher reflected the 12% chargeability, and 0.2% copper reflected a 5% chargeability, then the 12% chargeability contour could contain 5–10 million tonnes of ore assuming an average thickness of 30 m. If a portion of the chalcocite blanket with a lower content of, say 0.2%, copper, along the contour of the IP anomaly at 5% is included, then this would make it approximately 50 million tonnes of ore. These numbers were significantly lower than the requirements for the minimum volume of mineralization of interest to BHP.

At South Oyu Tolgoi, a rough calculation of the volume of mineralization was estimated to be 100 million tonnes at 1% copper taking into account that the mineralized zone extended in a west-north-westerly to northwesterly direction and assuming 200 m extension to depth. If the zone of oxidation, which can be processed by heap leaching, were to be included, then an additional 30 to 40 million tonnes of low-grade ore grading 0.25%–0.30% copper content could be added.

Considering a potential dilution factor, we estimated that to a depth of 200 m, we could hope for about 120 million tonnes. Similarly, to South Oyu, the South-West Oyu prospect standing alone did not meet the requirements of BHP either, and therefore, it had to be also considered in conjunction with all other prospects of Oyu Tolgoi. These estimates, which are likely best-case scenarios, indicate that it is unlikely that any one of the prospects would satisfy BHP's ore volume requirements on their own. Oyu Tolgoi would only be large enough for BHP if at least a few of the prospects coalesce into one major deposit.

Nevertheless, preliminary results of the first stage of drilling were encouraging. The unanimous opinion among the geologists is that we should conduct more drilling at Central and South Oyu Tolgoi, as well as on the southern flank of the Southwest Oyu.

Drilling continued until the first frost. By virtue of the impending winter cooling, the General Manager made a decision to stop drilling and resume it next spring (Fig. 4.15).

FIGURE 4.15 Oyu Tolgoi Ovoo. *From left to right: S. Diakov, G. Oyun, S. Myagmar, D. Garamjav, two drillers from Gobi Geo, S. Sanjdorj.*

4.5 DRILLING FOLLOW-UP PLANS

Based on the positive results obtained during the 1997 field campaign, BHP approved the proposal presented by the Oyu Tolgoi team to continue the exploration program in 1998. The following recommendations were made for the 1998 field program:

- conduct a transient electromagnetic imaging system ground test used effectively by BHP at Escondida for detection of and mapping the extensions of chalcocite blankets. If effective, consider covering large areas of prospective ground with airborne transient electromagnetic imaging;
- cover the entire area with a geochemical BLEG survey (nearly 200 samples);
- follow-up of the Khokh Khad and Shivee Ovoo prospects with geological mapping in conjunction with IP and magnetic surveys;
- continue studying mineralogical composition of the Oyu Tolgoi ore to clarify various metallurgical aspects of mineralization of different types for processing;
- conduct a second phase of drilling with 16 vertical and inclined diamond drill holes to do further and deeper testing at Central Oyu Tolgoi, South Oyu Tolgoi, on the southern flank of South-West Oyu Tolgoi and to infill drill the IP anomaly. The pattern of planned holes would also indicate the potential for coalescence of two or more of the prospects.

Sites proposed for additional drilling are on the map of anomalies of IP in green (Fig. 4.16). Consequently, we modified this proposal.

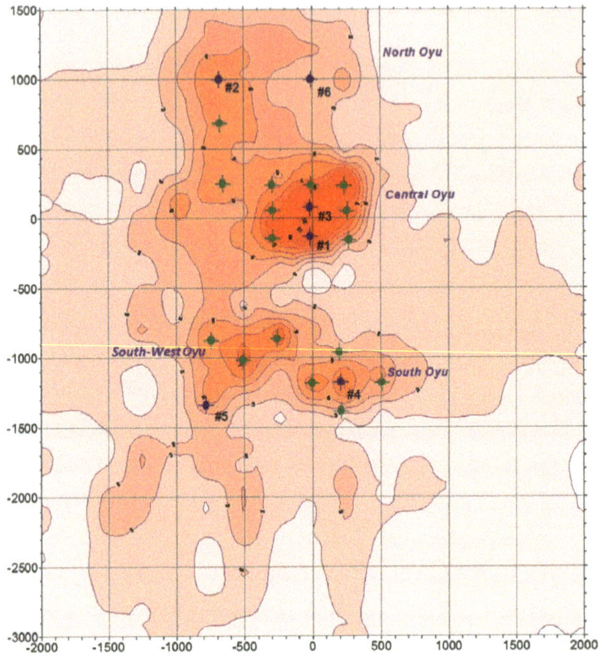

FIGURE 4.16 Position of drill holes suggested for drilling during field season 1998 (shown in green). The background is IP chargeability.

Plans for the increased amount of work in 1998 required some additions to the local BHP team in Mongolia. Sergei Diakov invited Barrie Bolton as a manager of BHP exploration program in Mongolia. Barrie, with his wife Helen, relocated from London to Ulaanbaatar. In addition, D. Garamjav was appointed to the permanent position of senior geologist and. S. Sanjdorj became the new Office Manager.

4.6 SECOND STAGE OF EXPLORATION

In 1998, the geologists started routinely recording magnetic measurements of the rock outcrops and mineralized zones during the course of mapping or prospecting and of the drill core.

During the 1998 field season, an airborne magnetic survey was flown over the Oyu Tolgoi license area consisting of 2,000 linear kilometers. The survey identified a number of additional magnetic anomalies and provided further clarification of the structures related to the deposits. Despite this survey did not help to find more ore at Oyu Tolgoi, it was helpful in structural interpretations on the district scale.

Originally, an airborne Transient Electromagnetic Method survey had been proposed. BHP had extensive experience with this method in Chile and found that it was effective in tracing the distribution of chalcocite blankets. However, this survey was not implemented.

For the second stage of drilling, it was essential to improve the quality of drilling, particularly productivity and the increase in core recovery. At that time, new drilling companies with adequate drilling equipment started arriving in Mongolia, such as Gobi Drilling Ltd., Can-Asia, and others. Ultimately, on April 14, 1998, Can-Asia won the contract for a 2,250-m drilling program.

Improvements in quality control for geochemical sampling were implemented by adopting BHP standards at Oyu Tolgoi. For each batch of 20 samples, it was required to insert one standard sample so that the result from the laboratory could be compared with the standard. This was meant to be a method for monitoring the quality of the laboratory and the quality of sampling method.

The overall conclusion among the BHP copper experts was that the Oyu Tolgoi drilling program was a technical success and that the BHP-Mongolian team had a good approach for conducting geological exploration, which was exemplified with the discovery of the chalcocite blanket at the newly discovered Central Oyu Tolgoi copper-porphyry system. Although chalcocite blanket was not large to satisfy BHP, we believed that it was very conceivable that a much larger supergene enrichment zone of copper mineralization was concealed under cover. It was equally conceivable that more and larger gold-enriched hypogene copper mineralization bodies also could be found, similar to the one intersected in borehole OT04 at South Oyu Tolgoi.

Furthermore, at Central Oyu Tolgoi, surface rock-chip sampling returned elevated arsenic suggesting that there is potential for enargite veining at depths probably not tested with the shallow phase 1 drilling. This would increase the chance for elevated copper content in the chalcocite blanket.

The field seasons with the second stage of drilling began on April 8 and continued until June 24, 1998. We drilled 13 drill holes with total amount of drilling reaching almost 2,128 m (Table 4.2). Following this phase of drilling, there were sufficient data to make cross-sections.

The core logging underpinned creation of cross-sections of 1:2,000 scale. Subsequently, J. Perelló together with S. Sanjdorj and D. Garamjav revised the core logs to produce a series of cross-sections, which we consequently used for resource calculation.

During the second phase of drilling, it was determined that zones with enhanced gold tenor were structurally controlled. Drill hole OT07 at South Oyu with an average content of 0.5% copper and 0.34 g/t gold included two assays with up to 11 g/t gold. Higher gold values were found within an interval of medium-grained feldspar-hornblende porphyries, which may represent the porphyry intrusion that generated the primary mineralization. Drill hole OT19 intercepted similar porphyry intrusion. However, these were devoid of sulfide mineralization and, most likely, represented a later phase of magmatic activity. In both cases, intrusive rocks were exposed to magnetite destruction resulting

TABLE 4.2 Summary Results of Copper Drilling in Drill Holes OT07–OT19

No. of Drill Hole and Location	Total Depth (m)	Major Mineralized Interval (m)	Intervals Maximum Content of Copper (%)
OT-7 South Oyu	280	164–190	1.2
OT-8 South Oyu	100	Barren	–
OT-9 South-West Oyu	251.9	114–200	1.1
OT-10 South-West Oyu	152.4	116–152	1.6
OT-11 South-West Oyu	100	Barren	–
OT-12 South-West Oyu	242.4	126–142.4	0.9
OT-13 Central Oyu	200.8	48–60	1.3
OT-14 North Oyu	145	120–145	0.85
OT-15 East Central Oyu	100.9	Barren	–
OT-16 West Central Oyu	76.2	Barren	–
OT-17 NW Central Oyu	120.7	62–76	0.45
OT-18 SE Central Oyu	200.25	178–188	2.4
OT-19 West South Oyu	157.29	130–150	0.55
Total meters	2,127.84		

in sericite–illite–smectite alteration assemblages, which correlated well with two magnetic low anomalies at South Oyu Tolgoi. Magnetic destruction also occurred along the northeasterly fault along which the porphyry intrusion was emplaced.

Drill hole OT10, drilled with the "golden triangle" at South-West Oyu Tolgoi, yielded gold values in excess of 1 g/t in places. Unfortunately, the drill hole had to be abandoned due to groundwater problems. The idea of twinning this hole was considered by moving the collar of the drill hole away from the tectonic zone. At that time, it was considered a priority largely because the new collar location would be along the OT09–OT10 where the potential for higher copper grades had been indicated. However, the recommendation was never implemented.

Based on the results of the second phase of drilling in 1998, the following conclusions were drawn:

- The Oyu Tolgoi resource increased. A few locations had potential to increase the amount of mineralization due to higher copper grades, particularly along the OT09 and OT04 profile, as well as copper–gold mineralization near drill hole OT10.

- Based on the drill information, mineralization on the western flanks of Central, South, and North Oyu Tolgoi from the western flank was limited due to the presence of a major granosyenite intrusion.
- An intrusion occupied the uplifted block along the tectonic fault of sub-longitudinal orientation.
- Postmineral dikes in drill holes OT10 and OT18 cross-cut the zone of mineralization with steep angles dipping at 55–60 degrees to the north.
- The Central Oyu Tolgoi zone of brecciation, intercepted by drill holes OT14 and OT18, required further exploration due to a relatively high copper content, grading up to 2.35% copper.
- Within Southwest Oyu, despite relatively low copper grades, there was a higher gold content of about 0.6 g/t and molybdenum of about 0.004%, which increased chances for better grades at the Oyu Tolgoi deposits.
- Numerous dikes of andesite, andesite–dacite, and rhyolite, particularly on the northern flank of the Southwest Oyu, confirmed the dynamic nature of the volcanic center, which probably lead to the hydrothermal activity responsible for the formation of Oyu Tolgoi. This is a typical feature of many copper porphyry systems.
- Andesite tuffs and tuff breccias intercepted in drill holes OT18 and OT15 suggested that the mineralization previously intercepted in drill hole OT02, might be somewhere nearby.

Promising results of the second stage of drilling indicated that the third stage of drilling would be needed to test the concept of a possible expansion of higher copper–gold grade zones of mineralization. This would require some deeper drilling near borehole OT10 where up to 3.04 g/t gold and 1.1% copper were found and also near drill hole OT04 where up to 3.4% copper was intersected.

4.7 AGE DETERMINATION OF OYU TOLGOI ROCKS

A potassium–argon (K–Ar) age measurement was made by Amdel laboratory in Australia of a copper mineralized potassic metasomatic core sample taken 75-m downhole in borehole OT09. Biotite from the sample gave the age of 411 ± 3 Ma with interpretation, which is the age corresponding to the potassic alteration event and associated copper mineralization. This age date is on the boundary between the Silurian and Devonian periods, making this the oldest copper porphyry deposit in Mongolia (Lamb and Cox, 1998), and possibly in all of Central Asia. This date was consistent with the early Paleozoic age previously proposed by the Mongolian geologists from the Burenhuu party (Burenhuu et al., 1995). This differed from the previously suggested Carboniferous age based on dating of fauna in the volcanics (Cox et al., 1997).

Another sample was from the surface within Central Oyu Tolgoi of an alunite crystal from a vein. The result gave absolute age of 117 ± 1 Ma, which was likely to correspond to the age of supergene oxidation and formation of the zone of secondary sulfide enrichment at Central Oyu. An age date of 117 ± 1 Ma was obtained, which corresponded to Early Cretaceous falling in the border of the Barremian and Aptian stages of the Cretaceous period. Since the sample was from the surface, it probably provided maximum age of chalcocite formation and the leaching process. This was in good agreement with the regional data, including the remains of dinosaurs in the Gobi desert.

4.8 THIRD, FINAL STAGE OF DRILLING AT OYU TOLGOI BY BHP

The third and last stage of drilling by the BHP team was conducted during September and October 1998. Initially, the proposal to drill nine holes totaling 2,250 m was made. However, these plans were not completed. Instead of the planned nine drill holes, only four, namely OT20–OT23, were completed totaling 761 m. All of the mineralized intervals in the Oyu Tolgoi drill holes are presented in Table 4.3.

The holes were drilled based on the recommendations from the geochemists and geophysicists after completing the airborne magnetic and soil geochemical surveys in 1998. The low magnetic anomalies on the southern flank of Central Oyu were a key drill target to determine whether they reflected the continuation of the chalcocite blanket. Drill testing was also proposed to resolve the magnetic anomalies in the south, which supposedly indicated the presence of a zone of magnetite-bearing potassic metasomatism, which may host higher copper, and gold grades similar to the previously intercepted mineralization in OT04 (see Fig. 4.17).

The third-stage drilling results did not deliver outstanding mineralized intercepts. Three out of the four drill holes were barren. Drill hole OT23 of South Oyu intercepted a relatively weak zone of mineralization with about 3% sulfides averaging 0.34% copper and 0.13 g/t gold in a 24-m interval from 196.8 to 220.8 m. Most of the mineralization occurred in quartz–chalcopyrite–bornite veins hosted by andesitic tuffs with chloritic alteration and disseminated magnetite, pyrite, chalcopyrite mineralization.

After completion of the third stage of drilling at Oyu Tolgoi, BHP had drilled a total of 3,789 m in 23 drill holes (Table 4.3) within a 5 square kilometers area. The spacing between the holes was about 400 m. The longest hole was 280 m and the average length was 165 m. Eight of the 23 drill holes intersected significant intervals of mineralization. It should be stated that at the end of BHP's drilling campaigns, the mineralization that had been intercepted by drilling had not been fully closed off to the south and west.

TABLE 4.3 Summary of Mineralized Intervals in Boreholes Drilled by BHP at Oyu Tolgoi

Drill Holes	Depth, м	Mineralized Intervals from - to, м	Mineralized Intervals, м	Average Grades		
				Copper, %	Gold, g/t	Molybdenum, %
OT01	136	26.0–52.0	26	0.86	0.07	0.004
OT02	98.9	72.0–86.0	14	1.09	0.23	0.002
OT03	186.3	18.0–56.4	38.4	1.63	0.07	0.003
OT04	250.7	54.0–126.0	72	1.61	0.18	0.001
		191.5–212.5	11	0.78	0.07	0.003
OT05	208	82.0–224.0	140	0.56	0.46	0.019
OT06	121.4			Barren		
OT07	280	108.0–137.4	29.4	0.36	0.3	0.0003
		151.6–190.0	38.4	0.7	0.29	0.002
		204.0–212.3	8.3	0.4	3.27	0.0015
		223.7–277.7	54	0.4	0.8	0.0005
OT08	100			Barren		
OT09	242.9	24.0–242.9	218.9	0.5	0.4	0.0093
OT10	152.4	30.0–42.0	12	0.6	0.27	0.017
		113.5–150.0	36.5	0.69	1.61	0.0021

Continued

TABLE 4.3 Summary of Mineralized Intervals in Boreholes Drilled by BHP at Oyu Tolgoi—cont'd

Drill Holes	Depth, м	Mineralized Intervals from - to, м	Mineralized Intervals, м	Average Grades		
				Copper, %	Gold, g/t	Molybdenum, %
OT11	100				Barren	
		56.0–72.0	16	0.39	0.07	0.0096
OT12	142.4	100.0–117.2	17.2	0.41	0.04	0.0066
		123.1–142.4	19.3	0.68	0.22	0.0019
OT13	200.8	46.0–90.0	44	0.37	0.14	0.0151
OT14	145.1	120.0–145.1	25.1	0.54	0.07	0.0144
OT15	100.9				Barren	
OT16	76.2				Barren	
OT17	120.7	52.0–77.0	25	0.34	0.37	0.0022
OT18	200.3	174.0–190.0	16	0.29	0.06	0.0428
OT19	157.3	129.0–157.3	28.3	0.32	0.02	0.0045
OT20	250.3				Barren	
OT21	125				Barren	
OT22	157.1				Barren	
OT23	236.1	196.8–220.8	24	0.34	0.13	

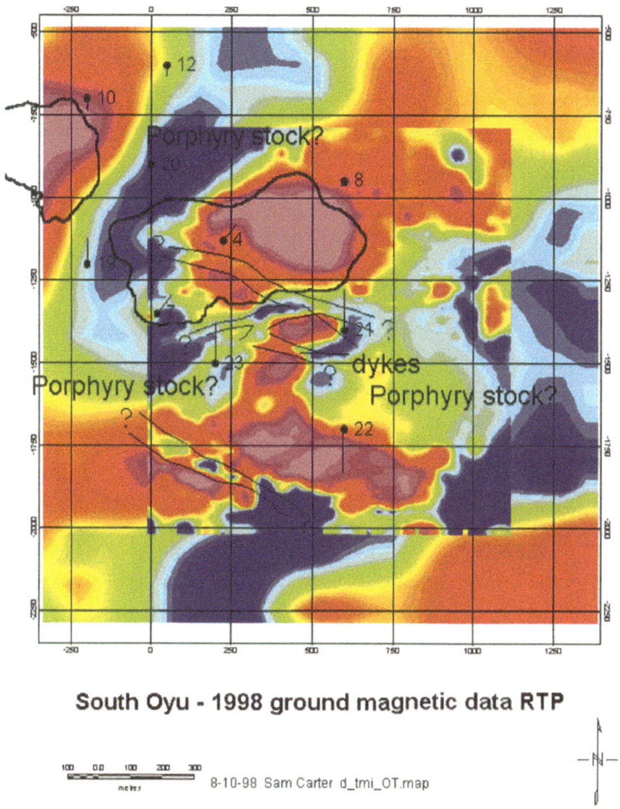

FIGURE 4.17 Detailed map of magnetic anomalies at South Oyu with the third-stage drill hole locations.

The concept of finding a porphyry system with chalcocite enrichment worked very well and was worth pursuing in other parts of the Gobi. However, the concept of high-grade hypogene mineralization failed to reach full testing and resolution at Oyu Tolgoi for several reasons. The company wanted to find a second "Escondida" with the orebody for open pit mining. Therefore, drilling at copper porphyry sites went mostly to a depth of 400 m with a step-out drilling distance no less than 400 m to make the right size of the orebody. Second, priority for copper porphyry exploration went to those copper porphyry prospects, which gave a better chance to find resources with high copper content in the zone of secondary sulfide enrichment (Fig. 4.18).

In fact, the existence of such blankets at the top of the orebody gave some winning moments. Increased copper content in the initial stage of mining significantly improved financial model of the project. In addition, copper in the zone of secondary sulfide mineralogy enrichment gave flexibility in technology

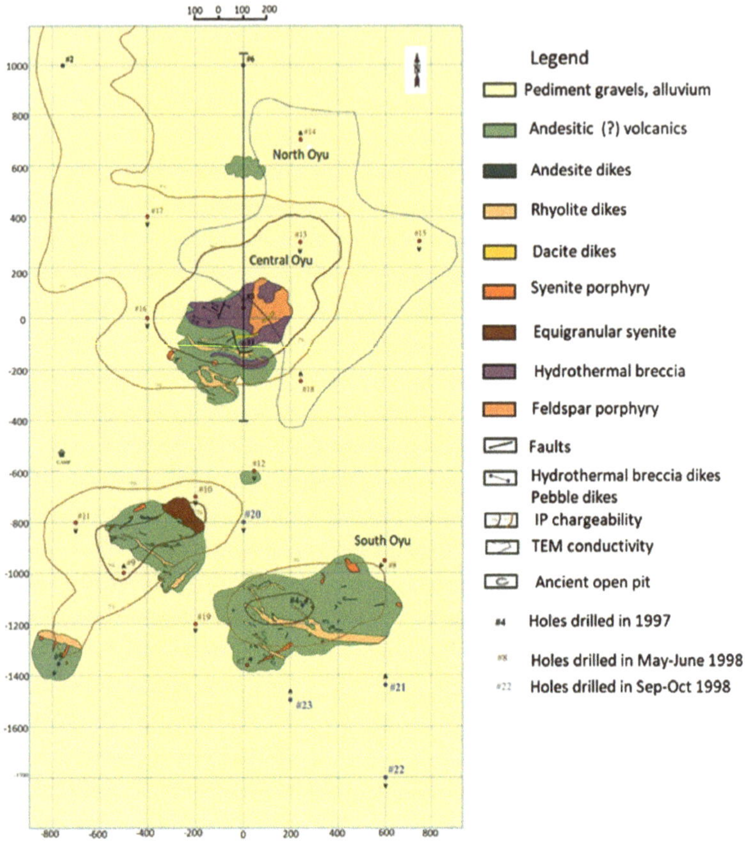

FIGURE 4.18 Map of Oyu Tolgoi BHP drill hole locations as of 1998.

for copper extraction. Certainly, many of these elements transpired from Escondida mine in Chile. We must say that Escondida is still the flagship not only for Copper BHP but also for the entire global copper industry. However, narrowly focusing on the Chilean model of copper porphyry exploration by BHP had played a "disservice" to Oyu Tolgoi at its initial stage. This prevented us from assessing fully the upside potential value of the hypogene mineralization at the deposits of Oyu Tolgoi.

4.9 MODELING AND CALCULATION OF EXPLORED RESOURCE

Based on the mineralized intersections from the 23 drill holes, we attempted to estimate the volume of the Oyu Tolgoi copper, gold, and molybdenum resources. Calculations were made in accordance with the JORC system of reserve definitions; no metallurgical studies were conducted, therefore no metallurgical considerations were made at this stage. To facilitate the resource calculations, the geological cross-sections produced by Garamjav, S. Sanjdorj, and J. Perello were used. J. Perelló was given the responsibility for resource delineation and calculation.

Christopher Ford, who like J. Perelló participated in the early stage of exploration at the Escondida deposit, processed the information and constructed a computer-based geological model using the GemCom program.

The conditions underlying the calculation of the resource included the following:

- The boundaries of mineralization were drawn out in sections based on visual assessment zones with similar content, which often coincided with the lithological boundaries, for example, in case of postmineral dikes.
- The calculation only included zones of mineralization with a minimum interval of 6 m (three contiguous 2-m intervals). Based on the statistical distribution of copper grades, classes of copper mineralization, and sets of economic limits were established.
- Anomalously high grades, such as a single 2-m interval with 12 g/t gold, for example, were smoothed to an interval of 8-m, while the average content of 3 g/t was reduced to 1 g/t.
- A series of bench plans at a scale of 1:2,000 based on the cross-sections were drafted.

All the surface mapping, core logging, assay data were then digitized and converted into a 3D mineral resource model using GemCom computer software. GemCom generated 3D polygons around the mineralized intervals, and with the help of the software, we calculated the volume of each polygon using the average weighted grades.

The resource calculations for each prospect area are presented in Table 4.4. In summary, Oyu Tolgoi as previously mentioned had an estimated resource of 438 million tonnes of ore with an average grade 0.52% copper and 0.25 g/t gold. The copper equivalent, considering only gold, was 438 million tonnes with an average grade of 0.67% copper. At that time, these numbers would have positioned Oyu Tolgoi in the top 40 undeveloped copper deposits worldwide.

It is noteworthy to emphasize again that the intercepted mineralization was not fully closed off to the south and the west. If the mineralization continued in these directions, then from 245 million tonnes to 900 million additional tonnes of ore could be located here down to a depth of 260 m. Moreover, if the mineralization is extended down to 460 m, the resource volume could double. An unverified potential of mineralization extending north to northeastward from Central Oyu was not taken into account for this resource calculation.

At the beginning of 1999, BHP estimated that the Oyu Tolgoi porphyry deposit had a potential resource of up to 1.2 billion tonnes of copper–

TABLE 4.4 Results of the 1999 Preliminary Resource Calculation at Oyu Tolgoi

Orebody	Mineralization Type	Tonnage, Mt	Copper Grade, %	Gold Grade, g/t	Molybdenum Grade, %
North Oyu	Primary	8	0.53	0.06	0.01
NW Oyu	Primary	10	1.09	0.43	0.002
Central Oyu	Secondary	10	1.1	0.1	0.02
Central Oyu	Primary	80	0.51	0.08	0.01
South and SW Oyu	Primary	331	0.48	0.3	0.015
Total		438	0.52	0.25	0.012

gold–molybdenum. Based on these figures, Oyu Tolgoi would be the second largest copper porphyry deposit in Mongolia with its fully unassessed size and grade potential; it would rival Erdenetiin Ovoo, the largest mine in Mongolia at that time.

In summary, it may be concluded that by conducting exploration along the guidelines of the copper porphyry model, the Oyu Tolgoi exploration team was successful in fulfilling the objective of discovering a copper porphyry deposit with the supergene enrichment component. However, the drilling strategy pursued by BHP at Oyu Tolgoi was designed specifically to delineate a shallow supergene enriched copper porphyry deposit, similarly to BHP's Escondida mine, that could be mined by the open pit method. Initially this strategy appeared to be adequate because at Central Oyu Tolgoi a classic high-grade supergene enriched blanket was discovered. However, with each successive drill hole, it became more and more apparent that the bulk of the mineral wealth at Oyu Tolgoi would not be near surface, but rather in the deeper-seated hypogene zone of the porphyry copper system. The latter style of deposit was not amenable to exclusive open pit extraction. Because BHP's intent was to find a second Escondida with a footprint of at least a few square kilometers, drill spacings as wide as 400 m were adequate and drill depths could be limited to 400 m or less. However, the length of the drill holes was insufficient, and the drill spacing pattern was too wide to delineate the hypogene zone at Oyu Tolgoi.

It must be stated that the financial model for supergene deposits is superior because supergene mineralization is typically of higher grade and less costly to refine than hypogene mineralization and open pit mines are much less costly operations than underground mines. It is likely that BHP stubbornly adhered to the supergene exploration model because they calculated that with the lack of infrastructure and local expertise in Mongolia at that time the expense of underground mining would be prohibitive. Narrowly focusing on the Chilean model of copper porphyry exploration by BHP was a "disservice" to Oyu Tolgoi at its initial stage. Because of these strategies, the BHP exploration team was unable to fully explore Oyu Tolgoi and assess its full value potential.

Chapter 5

Mongolian Government Support and Oyu Tolgoi Discovery Claim

After successful drilling at Oyu Tolgoi, the country's political leaders, including members of the Great Khural (Mongolian Parliament) recognized the exploration advances BHP had made in southern Mongolia. In August 1998, a BHP delegation, led by Hugo Dummett, was received by the Prime Minister of Mongolia (now the former President of Mongolia), Mr. Tsakhiagiin Elbegdorj (Figs. 5.1 and 5.2). The Leader of the Government expressed his interest in the Oyu Tolgoi program and Hugo Dummett reciprocated by giving a thorough presentation of the progress at Oyu Tolgoi. Hugo shared his optimism for the discovery of a large copper and gold deposit in southern Mongolia and BHP's desire to explore future opportunities for major

FIGURE 5.1 BHP delegation in front of the Government House of Mongolia. From left to right: Barrie Bolton, Sergei Diakov, Nick Allen, Hugo Dummett, Don Schissel, and Eric Seedorff.

Discovery of Oyu Tolgoi. https://doi.org/10.1016/B978-0-12-816089-3.00005-6

FIGURE 5.2 Meeting of BHP delegation with the Prime Minister of Mongolia Tsakhiagiin Elbegdorj (formerly elected President of Mongolia).

investments in Mongolia to further advance the development of the mineral and energy sectors of the country.

By this time, the evaluation of Tavan Tolgoi coalfield located at a distance of 120 km from Oyu Tolgoi was advancing, and in the case of a positive outcome for both Oyu Tolgoi and Tavan Tolgoi, BHP intended to submit detailed proposals for these two projects to BHP management. To proceed, it was essential that the stability of the investment climate in Mongolia was on a solid base. The Prime Minister ensured BHP that the operation environment would be favorable for carrying out the necessary work and wished the company success in the achievement of its objectives in Mongolia.

Members of the Oyu Tolgoi team have fond memories of an unforgettable meeting with Member of Parliament Luvsanvandangiin Bold, who was interested in the work of the BHP in the Southern Gobi. He was shown a few samples of drill core from OT04. The most impressive was a short interval with bright native copper veinlets. Mr. Bold's question was what did it mean? The answer sounded like this: "the preliminary result obtained at Oyu Tolgoi indicate that in case of a positive outcome, we are dealing with a potentially large deposit of copper and gold. If all our assumptions come true, then we can conclude that in the long term the center of Mongolia's copper industry would gradually move to the South Gobi. Southern Gobi will be the future of the Mongolian copper production. Of course, all of this would require infrastructure development. Nevertheless, the day was fast approaching when Boeing 737 aircraft would start flying regularly to the Gobi." At that time, this prediction seemed incredibly bold, like something out of science fiction. Now we see how that future vision is already becoming a reality and part of daily life for the inhabitants of the Gobi Desert in Mongolia. There is a giant copper mine currently operating in the Gobi and the Oyu Tolgoi employees are coming to work on a fly-in-fly-out basis by Boeing 737 aircraft.

The efforts of BHP received broad support from the leading experts of the mining industry in Mongolia. It was especially heartening to hear expressions of genuine interest from the first President of Mongolia, His Excellency Punsalmaagiin Ochirbat, who himself was a specialist in mining and had a doctorate of technical sciences. Mr. Ochirbat visited the BHP Office in London to inquire into the progress made at Oyu Tolgoi. He subsequently also visited the Oyu Tolgoi deposit. His memories of the events leading up to the discovery of Oyu Tolgoi are outlined in his book "Oyu Tolgoi. Past, present and future" (Ochirbat, 2010). During an investment conference in Ulaanbaatar "Discover Mongolia" in September 2005, he expressed full support for the efforts of both national and international mining companies engaged in prospecting and exploration of mineral deposits in Mongolia (Fig. 5.3).

After the second stage of drilling, it became clear that in southern Mongolia, BHP made a significant discovery. The company prepared and submitted a letter to the Head of the Agency for Mineral Resources Mr. Jargalsaikhan requesting registering discovery of the Tolgoi Oyu deposit. The letter highlighted that BHP, while performing exploration work for copper and gold in the Gobi, discovered a new deposit in Mongolia named Oyu Tolgoi.

The Oyu Tolgoi discoverers list included Dondogiin Garamjav, Dennis Cox, Samand Sanjdorj, Sergei Diakov, Tumur-Ochiriin Munkhbat, and Sam Carter. The list was accepted, and in the response letter, the Agency promised to prepare discoverer certificates for all participants.

FIGURE 5.3 At the conference « Discover Mongolia » Ulaanbaatar, 2005. From right to left: President P. Ochirbat, A. Diakov, and S. Diakov.

Chapter 6

Regional Prospecting for Copper and Gold

The success at Oyu Tolgoi gave an additional boost to BHP regional generative program in Mongolia focusing on copper and gold. In 1997 the Mongolian branch of BHP was awarded the Berkh, Rashant, Khubsugul, and Zamtiin Khad exploration licenses advancing its copper porphyry program in Dornogovi aimag (Table 6.1 and Fig. 3.8). During 1998 field season, BHP conducted reconnaissance work in these areas. D. Cox led the team working in the new areas, while the field activities at Oyu Tolgoi and Ih Shankh were guided by S. Sanjdorj.

In addition to 1:25,000 scale mapping on the Khubsugul license, received in June 1997, the field team conducted geochemical rock chip sampling of primary bedrock and overlying cover rocks.

On the Berkh license, received on September 22, 1997, BHP carried out selective mapping and geochemical sampling. Based on the results of the fieldwork, it was concluded that the area had limited potential for copper. The southeastern part of the license area was composed of Riphean gneisses with numerous quartz veins and lenses. The team collected 155 BLEG samples, which returned several gold values ranging from 1.1 to 1.6 ppb and a maximum silver value of 59 ppb. A 1.55 g/t gold result came from one of the geochemical rock chip samples while a few other samples yielded elevated contents of arsenic, antimony, mercury, and barium, but copper and molybdenum were at the background level. It was recommended to reduce the license area to 727.75 square kilometers and to conduct further exploration for gold.

Reconnaissance was conducted on both the Rashant and Zamtiin Khad licenses, received on September 22, 1997. On Rashant, it was recognized that the dominant intrusion of Permian age plagiogranite had limited potential for copper, and therefore, it was suggested to drop the license. The Zamtiin Khad was also received on September 22, 1997. After field visit, the Zamtiin Khad was reduced to 676.5 square kilometers and thoroughly explored in 1998.

Discovery of Oyu Tolgoi. https://doi.org/10.1016/B978-0-12-816089-3.00006-8

TABLE 6.1 List of Exploration Licenses for Copper in Regional Reconnaissance Program

License Area	Coordinates Northing	Coordinates Easting	Area, sq.km
Rashaant	44.0167	110.85	1,153.42
	44.1833	110.85	
	44.1833	111.5	
	44.0167	111.5	
Khubsugul	43.5667	109.5	525.93
	43.7	109.5	
	43.7	109.9667	
	43.5667	109.9667	
Zamtiin Khad	43.8167	110.9	1,105.98
	43.9167	110.9	
	43.9167	111.5	
	43.8167	111.5	
Berkh	43.6333	110.1667	2,214.92
	43.8167	110.1667	
	43.8167	111	
	43.6333	111	
Total			5,000.25

At the Ih Shankh license no. 211, which was granted on January 7, 1997 with an area of 1,168.5 square kilometers, the geoscientists rechecked the license area for signs of alteration. The team collected 15 BLEG samples and 266 geochemical rock chip samples from primary rocks. They also made more than 70 PIMA measurements of various types of altered rocks. Although the geochemical results were disappointing, some intriguing overlapping alteration phases were found in the West Gashuun portion of Ih Shankh. Exploration continued on the Ih Shankh license at West Gashuun covering an area of 46,750 ha. Part of the area was geologically mapped at a scale of 1:5,000. In the spring of 1999, we drilled two holes with a total of 501 linear meters of diamond drilling. Drill hole IS01 intersected weak chalcopyrite in intervals 120.4−135.0, 179.2−185.8, and 185.8−198 m associated with quartz-sericite alteration and hydrothermal breccias in andesite and dacite porphyry. In drill hole IS02, copper was found at a depth of 236 m in syenite granodiorite porphyries. In both drill holes, widely distributed disseminated pyrite was observed.

During the period from August to September of 1998, D. Garamjav and T. Munkhbat led the reconnaissance copper exploration in the eastern and western extensions of the Late Devonian to Carboniferous age Gobi volcanic belt. Here the focus was the known prospects of Bogd, Khurmen, and Mandakh. All these prospects contained numerous copper showings.

Concurrently with the regional copper exploration program in 1998, BHP also conducted regional prospecting for gold. BHP's minimum size requirement for this commodity at least 6 million ounces of gold was a significant size of gold deposit not yet known in Mongolia. This would be a challenging task. To achieve this goal, the team knew that it would have to focus on geological environments that could produce a world-class gold deposit. Based on the team's knowledge of Mongolian geology, they believed that the two types of gold mineralization that could potentially be found were Carlin-style interbedded limestones, similar to those identified in Nevada, and also black shale-hosted gold deposits similar to the giant Muruntau deposit or Sukhoi Log.

B. Bolton and G. Jamsrandorj conducted research for the availability of such environments in the Precambrian and Early Paleozoic sediments of Mongolia. They identified five areas with the development of black-shale sequences. With the aid of the Mongolian database and 1:1,250,000 scale gold mineralization maps, the geological settings with Carlin and Muruntau-Sukhoi Log styles potential were selected and ranked. The highest ranked areas appeared to be the Bayankhongor and Khubsugul regions. Jamsrandorj strongly suggested the latter because it appeared to be a favorable environment for both deposit types.

The BHP field team, led by G. Jamsrandorj, evaluated the potential for Carlin-style deposition on the Delgerekh license area. BHP Senior Geologist, Mark Osterberg, an experienced gold explorer and known for his contribution to the discovery of a significant gold deposit in Nevada named Twin Creeks, joined the team. The Delgerekh license area had been geologically mapped at a scale of 1:200,000, compiled by J. Burentugs in 1990 and had 1:100,000 scale SPOT imagery. A 50 square kilometers area was mapped by BHP in the central part of the license at a scale of 1:10,000. During the process of mapping, 100-m-wide jasperoid bands stretching for a length of 25 km along the contact of limestones and quartz clastic sedimentary rocks. At three sites, Carlin-type alteration was seen. Evidence of tin, tungsten, and molybdenum mineralization was also found by the geologists. Jasperoid exposures with quartz stockworks were sampled by 250 geochemical rock chip samples from primary silicified rocks and 1,200 geochemical soil samples from secondary geochemical halos. Rock chip samples were sent to Ana-Labs in Ulaanbaatar for preprocessing. All soil samples were sent directly to the Chemex laboratory in Vancouver. The 150 BLEG samples were sent to Australia for analysis.

Mapping determined that at Delgerekh, 3,000- to 4,000-m-thick sequences of calcareous and clastic sedimentary rocks trended in an east-northeasterly direction (70 degrees azimuth) with a steep dip to the south. To the north, there was a granite intrusion, and the sedimentary sequence was metamorphosed to gneisses, schists, quartzites, phyllites, and marbles. About 90% of the carbonate column was composed of limestones and dolomites and the remainder by thin interbedded sandstones.

In four locations along the 30-km trend, the field team observed zones of pyritization and hematitization, sometimes with intense silicification and the development of the jasperoids at the top of the section. Two anomalies were drill-tested with two drill holes. The first drill hole intercepted two zones of mineralization with slightly elevated gold contents: 0.1−0.29 g/t from 9.5 to 50 m and less than 0.29 g/t from 99 to 106 m. Elevated mercury content (up to 2.58 ppm) and arsenic (up to 190 ppm) also occur. No further exploration was recommended on this license.

At Lake Khubsugul, the team also conducted the fieldwork. G. Jamsrandorj was assisted by Andy Wilde who was an experienced gold explorer from Australia. Near Bayan Zurkh, the field team mapped northwesterly structures and attempted to locate gold mineralization with geochemical indicators arsenic-tin-tungsten, and mercury. Here 184 BLEG samples were collected but the analysis was disappointing. The gold, silver, arsenic, and bismuth results were predominantly at background levels.

In 1998 the Government adopted a series of legislative measures that drastically reduced the attractiveness of the investment climate for gold exploration in Mongolia. During the fall session of the Great Khural on November 6, 1998, the Mongolian government introduced a 10% tax on gold exports from Mongolia. This unexpected development made gold mining much less attractive than in other jurisdictions. Therefore, BHP decided to suspend its gold exploration programs and surrender all its gold exploration licenses. BHP rehabilitated all areas to their original condition where any ground disturbance had occurred.

Chapter 7

Integrated Approach to Resource Development, Tavan Tolgoi

When exploration results at Oyu Tolgoi began to look promising, BHP management had a second look at infrastructure issues, transportation, water, and energy. The distance from Oyu Tolgoi to Sainshand railway station, the nearest railroad in Mongolia, was 340 km in a straight line. Alternative routes were also considered. One was a direct line in southwestern direction to Bayan Oboo across the border with the People's Republic of China in Inner Mongolia. The third option was a straight line to the south to the railway station Bayannur on the Chinese northern railroad line (see Fig. 7.1). Each route has its pros and cons to be discussed below. The one to Bayannur was the shortest but the one to Bayan Oboo in Inner Mongolia of People's Republic of China was simpler and less costly due to more flat topography, fewer river streams to cross and bridges to build, etc.

The source of an energy supply was another challenging problem. Considering the few available options, extension of the high-voltage power lines from China to the Oyu Tolgoi seemed to be the most expedient one. However, the preference of the Government of Mongolia was to develop its own sources of energy. If BHP were to comply with the Government's wishes, then the only practical option would be to develop resources in the Gobi Desert. The closest coal deposit to Oyu Tolgoi was the giant resource at Tavan Tolgoi, which was being mined only to supply local heating needs. The previous exploration conducted by the government of Mongolia in cooperation with the Soviet Union had outlined and calculated a multibillion tonne resource to a depth of 350 m. There was no doubt that a large coal reserves/resource was present. Development of a copper mine in conjunction with a coal deposit, from which coal can be used as a source of energy for copper production and also supplying metallurgical quality coking coal to the external markets, was an appealing concept to BHP. There were markets for coking coal in the neighboring Mongolia countries, Russian Federation, and People's

Discovery of Oyu Tolgoi. https://doi.org/10.1016/B978-0-12-816089-3.00007-X

FIGURE 7.1 Main directions of transport routes for shipping coal to the markets.

Republic of China. In addition, South Korean and Japan's metallurgical industries also were in great demand for good-quality coking coal. The question was whether it would be profitable to deliver the coal products from the deposit to these markets.

At this time, the Government had already contacted the World Bank to organize a tender for conducting a feasibility study for the development of Tavan Tolgoi. BHP, with its vast experience in coal mining in the Bowen Basin of eastern Australia and its global experience in marketing coal products, was an obvious candidate for this proposition. Both the World Bank and BHP responded positively. Furthermore, BHP indicated that it would be interested in financing the additional exploration and the feasibility study at Tavan Tolgoi and, if the results proved to be positive, it would be interested in the construction of a major energy complex in the southern part of the country. Attracting a large multicommodity mining company like BHP, which was also a copper producer actively involved in copper exploration in the Gobi Desert, to the development of Tavan Tolgoi coal deposit was an excellent solution to the issue of bringing large development projects and related infrastructure to a remote part of Mongolia.

The entire area of Tavan Tolgoi coalfields, except for a small postage stamp operation by a local mining company, was under the control of the State. BHP applied for two exploration licenses of about 200 square kilometers covering the coal basin of Tavan Tolgoi (see Fig. 7.2). The first step was to conduct a comprehensive review of reports and documentation available for Tavan Tolgoi with some help from its coal mining group in Kuala Lumpur. At the disposal of BHP, there were 10 detailed geological reports in more than 150 volumes, which were readily available in the government geological funds library. In conjunction with this a BHP delegation of coal experts visited Tavan Tolgoi from October 17 to 24, 1997.

The Permian age Tavan Tolgoi coalfield is located in the intermountain trough of the South Gobi belt. It had been explored during a 40-year long period by Bulgarian and Mongolian geologists. In the 1980s and 1990s, several drilling programs had been conducted totaling approximately 800 drill holes, most of them in the order of 250 m deep with drill spacings ranging from 150 to 350 m. The drill hole data were of a high standard consisting of specific density measurements, gamma-ray downhole logs, petrographic studies, and coal quality analysis that was conducted for all varieties of coal.

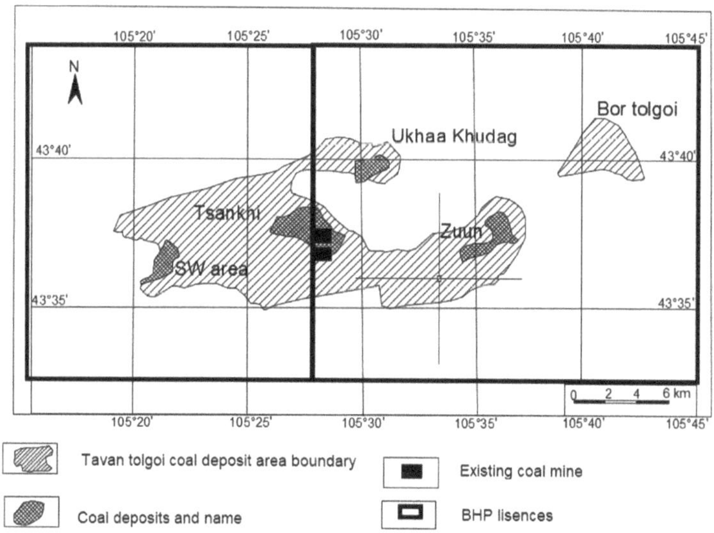

FIGURE 7.2 Boundaries of BHP exploration licenses at Tavan Tolgoi.

The Tavan Tolgoi stratigraphy is characterized by a rhythmic deposition of conglomerates, sandstones, mudstones, and coal. The mudstones contain a high amount of organic matter and acted as the marking horizons. The coal-bearing sequence was 1,000 m thick and consisted of 16 coal seams (labeled as 0 to XV in ascending order). The thickness of the individual beds ranged from 0.3 to 46.5 m. The thickest seams were VIII and IX. Structurally, the coalfield consisted of a series of synclines and anticlines, elongated in the latitudinal direction, on which an anticlinal fold of sublongitudinal strike was superimposed. Angles of dip on the wings reached 40 degrees. A series of sublatitudinal, discontinuous, tectonic normal faults and two major north-western and northeastern thrust faults crosscut the coalfield. This faulting presented potential mining risks due to structural complications.

Based on the contents of vitrinite, coal quality varied from 0.88% at the top of the sequence to 1.38% in its bottom part. Seams 0, III, IV, and IX carried basal beds of coking coal. The estimated coal resource/reserves at Tavan Tolgoi was 6 billion tonnes for all five sites proposed for development, namely Tsankhi, Ukhaa Khudag, Zuun, Southwest, and Bortolgoi. The largest Tsankhi and Ukhaa Khudag, located in the Central and Eastern parts of the coalfields, contained about 2.15 billion tonnes, including 806 million tonnes identified as coking coal. Extracting coal from these beds through open cast mining would have a stripping ratio of 4:1.

Initial studies suggested that the potential of Tavan Tolgoi (see aerial view of the basin on Fig. 7.3) would be of interest to BHP because if it were located near-existing infrastructure and near tidewater, it would have started production long ago.

FIGURE 7.3 Aerial view on coking coal pit of Tavan Tolgoi coal deposit (October 2005).

The previous studies indicated that Tavan Tolgoi coals possessed a good calorie index and relatively low sulfur. This was an affirmation that in addition to valuable coking coal, the thermal coal from the deposit could be used for electricity production to meet the energy demand from both the local consumers and the future copper production at Oyu Tolgoi. However, the data available from previous coal quality tests were deemed to be insufficient by the BHP coal group. A more detailed review of the quality of all coal seams was necessary, and the coking properties of coal also needed to be established. This would be an important part of the feasibility study. For this a variety of bulk samples would be taken and additional holes would have to be drilled. The coal properties would then be categorized in a standard coal industry format.

The feasibility study envisaged by BHP reviewing all risk factors associated with the development of Tavan Tolgoi. It was also necessary to assess the trends for the requirements of coking and thermal coal in the markets, as well as the local needs for electricity. The most sensitive factors that posed an increased risk for the economics of Tavan Tolgoi were as follows:

- the availability of coal markets;
- properties of the coal and their suitability for the requirements of the markets;
- the lack of railways, roads and basic infrastructure;
- investment stability guarantees;
- project economic viability in the long run.

All preliminary studies were expected to be completed at a cost of about US$500,000. In case of a positive outcome, it was proposed that a preliminary feasibility study of the development of the facility would require at least an additional financing in the same amount of about half a million dollars.

BHP's economic study of Tavan Tolgoi was expected to be completed in three phases. The first step was to study the concept of a feasibility study to determine whether the coal reserves meet BHP's production, technical, and business requirements. This first-phase study was collaboration among personnel from BHP's exploration, energy, coal, transport, marketing, and copper divisions. To adequately assess the value of Tavan Tolgoi for the market, BHP's marketing divisions in Hong Kong and Beijing were used. The marketing study included research of electricity consumers in the region, the status of the internal market needs for thermal coal, the international market situation for coking coal, and the future dynamics for various brands of coking coals.

To confirm geological conclusions, it was proposed to summarize all geological information and, as mentioned previously, to conduct a short program of additional drilling and bulk sampling for a more detailed study of thermal and coking coal properties at Tavan Tolgoi. For an assessment of potential mine development, it was proposed to develop a conceptual model

FIGURE 7.4 Mining of coal beds at Tavan Tolgoi. *Photo dated 2006.*

based on real data from BHP coal operations with similar parameters. Small local production was ongoing at the site with the coal product being shipped to China (see Figs. 7.4 and 7.5). It was also suggested to undertake a comparative analysis of coal production from other competitive producers of coal. So all in all, it was an exhaustive study that would allow making an adequate business decision.

Other first-phase issues included availability of labor resources for processing and transporting coal, safety, environmental protocols, and creating a financial model for the concurrent development of both Oyu Tolgoi and Tavan Tolgoi, which would highlight cost-saving synergies. BHP was obligated to provide a copy of the report to the Government.

In case of positive first-stage results, the plan was to proceed to the second stage of the Tavan Tolgoi study, which was essentially a preliminary feasibility study. The tasks at this stage included the following:

- digitization of existing and newly collected data to facilitate three-dimensional geological modeling using the Mincom software system. The plan was to evaluate multiple options;
- selective drilling of coal seams and selection of bulk sampling to further clarify the quality and other characteristics of the various coal types;
- more detailed study of possible railway routes connecting Oyu Tolgoi and Tavan Tolgoi projects with the existing railroad network;
- further comprehensive study of viable markets for coal products with regard to potential consumers and their long-term demand;
- initial examination of the construction of heat power plants;
- examining consumer networks at the local and national level.

(A)

(B)

FIGURE 7.5 Tavan Tolgoi coal production (above—mining of coal seam, below—transportation of coal by long-haul trucks). *Photo 2006.*

The results of the second-phase research were to be compiled in one single report with preliminary feasibility study conclusions and recommendations, which the company had to submit to the Government of Mongolia by the end of 1999.

In case of positive prefeasibility recommendations, BHP was committed to perform the required work for the preparation of a final feasibility study. In this document, all parameters of the deposit model had to be finalized. The results of the final feasibility study were to be presented to the Board of Directors of a selected financial institution for approval of bank financing. It was required that the feasibility study document be approved and funding be finalized by the end of the year 2000.

The detailed agreement stipulated all the conditions and obligations of the license holder. The cost of the research in three stages stood at US$ 1 million dollars to be fully funded by the company from internal resources. In case of negative results, BHP was obliged to return all exploration licenses to the Government of Mongolia along with the copies of the conducted research results.

Construction of a railroad would be a necessity, the cost of which could be shared by both the Oyu Tolgoi and the Tavan Tolgoi projects. Three viable routes were considered in three possible directions (Fig. 7.1). The first leg would be to connect the two mines with a 120-km-long railroad, which would have an elevation drop of 500 m. From there, several directions were in consideration: (1) eastward toward Sainshand station, (2) directly to the South toward the railroad station Bayannur in China, and (3) south-east toward Bayan Ovoo. The Bayannur option, although the shortest with a distance of 253 km, was unattractive because there were multiple crosses through numerous river channels. The most preferred was an option with a railway spur-line going toward Bayan Ovoo for a distance of 290 km with an elevation differential in topographic relief of 300 m. This line would cross 5 roads and 21 rivers going from Tavan Tolgoi through Oyu Tolgoi site and then continuing southeast to the Mongol-Chinese border and ending in Bayan Ovoo, China. The team accepted this option. Besides limited railroad capacity in China, there was also limited capacity for loading coal onto ships in Tianjin.

Regarding the deposit of thermal coals, the experts considered an option of building a thermal power station with a capacity of 5 MW and then increasing it to 11 MW. To reduce transport costs for hauling ash components, the expert suggested conducting enrichment of coking coal on the site.

A more detailed review of the quality of all coal seams was necessary, and the coking properties of coal also needed to be established. For this a variety of bulk samples would be taken and additional holes would have to be drilled at the end of 1998. The coal properties would then be categorized in standard coal industry format. The most significant parts of the Tavan Tolgoi coal basin were Tsankhi and Ukhaa Khudag. Tsankhi was selected as a reference for drill testing of the three main coal seams. To this end, the company completed drilled 428.47 m in six drill holes (four main and two additional drill holes) (see Table 7.1).

The field team left Ulaanbaatar on October 5, 1998 to test three coal beds at Tsankhi, which represented the most important part of the coal basin for resource calculation. The drilling contractor was Gobi Drilling, a wholly owned subsidiary of the Australian drilling company Radial Drilling. They used an RD1500 drill made in Australia. The sequence of drilling went from drill hole BHP-4 to drill hole BHP-1, and then at the end with drill holes BHP-1A and BHP-2A. Drilling was conducted in challenging weather conditions and by the end of the drilling, on October 21, it was snowing.

TABLE 7.1 Drill Holes Completed by BHP at Tsankhi

Drill Hole	BHP-1	BHP-1A	BHP-2	BHP-2A	BHP-3	BHP-4	Total
Depth, m	82.5	84.5	58.5	54.67	77.9	70.4	428.47
Coal-bed	0	0	IV		III	III	
Core recovery, %	99	98.2	70	91.3	86.8	88.7	

Based on existing and newly acquired data, a conceptual model for open pits in the Tsankhi and Ukhaa Khudag coalfields was made with a total production of 10 million tonnes capacity per year for 20 years. It was assumed that one-third of coal production would be directed to the needs of the market in China and the remaining two-thirds would supply to the Asian markets, mainly South Korea and Japan. The model assumed that it would require about 9 years to reach the optimal production of 10 million tonnes per year, starting with 1 million tonnes per year, gradually increasing the annual production to 3 million tonnes, then 6 million tonnes, and finally achieving the optimal level of production.

The BHP laboratory in Australia conducted the coking tests and the results were inspiring. The tests confirmed, unequivocally, the coking ability in coal seams 0, III, and IV. The conclusion was unambiguous—Tavan Tolgoi had vast reserves of high-quality coking and thermal coal.

At the end of 1998, the coal experts prepared a final report of their results. Unfortunately, the conclusions did not meet our expectations. Their verdict was negative due to the lack of infrastructure, limited market demand, and low world prices for coal at that time. Hence, the Tavan Tolgoi mine did not meet the investment requirements of BHP, and the company decided to stop work and return the exploration license to the Government of Mongolia. On February 11, 1999, BHP handed over the detailed report with all statistical data and model calculations of Tavan Tolgoi study to the Ministry of Agriculture and Industry, as well as Agency for Mineral Resources in Mongolia.

In retrospect, the exploration team believed that the coal group did a thorough job but they failed to look into the details of the market dynamics by taking a short-term view rather than a long-term outlook perspective. The exploration team was deeply disappointed with the results and the subsequent decision of the company to suspend further interest in Tavan Tolgoi. Consequently, when the economic situation changed due to higher coal prices, it was impractical to relaunch the program.

The lack of infrastructure was always the main problem blocking development of the mineral resources in South Gobi region of Mongolia. Commissioned by the Mongolian Government and the World Bank, the Norwest Corporation that is known for its experience in providing engineering services for coalfields conducted their assessment of Tavan Tolgoi in 1998. They concluded that the project would be cost-effective only if coking coal prices were in excess of US$50 per tonne FOB. The market conditions at that time were not beneficial for the development of the project, which was confirmed by BHP's analyses. It was disappointing that the coal group experts did not seriously consider future scenarios for coking coal market trends hence missing the opportunity to develop one of the most significant undeveloped coal deposits.

Subsequently, the infrastructure situation started to improve. Soon after these studies, in China a railway line was built by Jiuquan Steel Company with a capacity of transporting 8 million tonnes per year to ship coal from the Mongolian border to Jiuquan. Consequent infrastructure developments in the south would radically change the Tavan Tolgoi profitability calculations as well as the conclusions and recommendations of the coal team.

New possibilities for the development of the region would become a subject of further studies. We will witness more brave decisions on the issue of the infrastructure of the southern Gobi area. In the meantime, a unique opportunity for integrated development of the South Gobi region with two significant deposits of coal and copper by a large mining company was apparently missed.

Chapter 8

Corporate Changes in BHP

Paul M. Anderson *was born in Richland, Washington State, USA. In 1967, he graduated from university with a Bachelor's degree in mechanical engineering. In 1969, Paul graduated from the MBA School at Stanford University in California. Paul started his career at Boeing, moved to the Ford Company, and then to Pan Energy. At Pan Energy he rose to the post of Chairman of the Board of Directors, and the President and Chief Executive Officer of the company. From 1997 to 1998, Paul was President of Duke Energy. In December 1998, he took the position of CEO of BHP. During his time at BHP, the company made numerous changes. In 2001, Paul Anderson presided over a merger of BHP with Billiton Company. In 2002, he handed over the reins of the combined company BHP Billiton to Brian Gilbertson, his counterpart at Billiton.*

Donald Robert Argus *was born in Australia in 1938. He graduated from the Anglican Church school in Brisbane, Queensland. Donald has the title of an honorary doctorate from the Griffiths University, Monash University, and from the Queensland University. Early in his career, Don was a banker. He played critical role in the revival of the National Australian Bank in the late 80s. In 1996, Don Argus became a member of the Board of Directors of BHP. In 1999, he took the Chairman position replacing Jerry Ellis. Donald remained in this position until the year 2010.*

In 1998, BHP went through some significant changes, which had a substantial impact on BHP exploration program in Mongolia.

Discovery of Oyu Tolgoi. https://doi.org/10.1016/B978-0-12-816089-3.00008-1

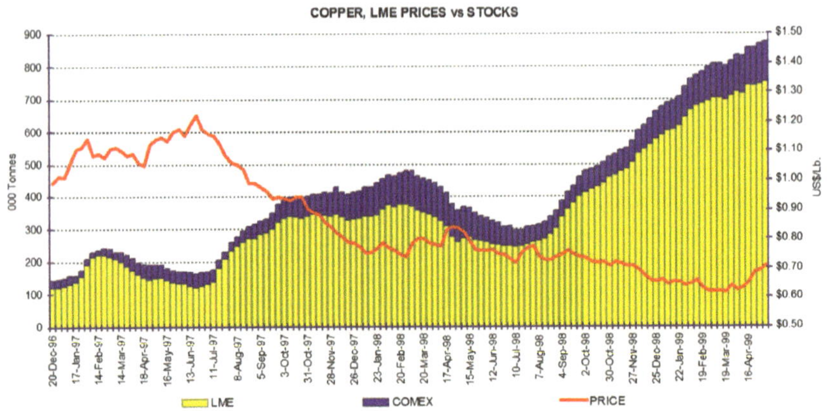

FIGURE 8.1 Trends in copper prices and copper supply to the global market in 1997–98 (LME historical prices).

Remarkably, all four BHP production subdivisions (minerals, copper, steel, and oil) suffered from a drop in global commodity prices. The North American copper division even had to suspend production at the mines in San Manuel and Robinson. When the copper price dropped to US$ 70 cents per pound (see Fig. 8.1), these mines were on the verge of profitability. Several options were under consideration. One was to continue the temporary suspension of production at San Manuel, mothball the Robinson mine but continue to operate a metallurgical plant in San Manuel by feeding the smelter with the concentrate from other mines. With this strategy, the technological chain could quickly return to the original production flow after the recovery of the copper prices. The second option was to close both mines and sell the smelter plant. There was a concern about the second option that in case of global commodity price recovery the company would lose access to the US market. Besides, the sale of the smelter at low commodity prices could hardly warrant any attractive deals. However, those were not all the challenges that the company was facing at that time.

After a series of setbacks with HBI (hot briquetted iron) plants, including the main plant in Port Hedland, Australia, the list of problems continued to grow. Because of falling copper prices, BHP's acquisition of Magma Copper at the end of 1996 started looking grim. The commissioning of the Beenup mining operation to extract titanium sands in Western Australia turned out to be marred with problems and there was also unpleasant news from Hartley, the platinum within the Great Dike in Zimbabwe, which started production in 1997. Immediately after the startup, it experienced numerous mining complications that prevented the mine from achieving its production goals. All these issues "snowballed" on BHP, all at once.

BHP shareholders became disenchanted with the top leadership, especially after publishing the company 1997 financial results, indicating that the company's profit was negative and BHP suffered substantial shortfalls.

The "red" list, published in May 1998, included the following losses: US$ 620 million dollars at the metallurgical plant in San Manuel, US$ 868 million dollars from the Arizona copper mines, and US$ 129 million dollars from the Tintaya mine in Peru. It also included US$ 378 million dollars from the Boodarie HBI production plant in Port Hedland, US$ 357 million dollars from Hartley, and US$ 99 million dollars from Beenup. Previously, BHP had suffered significant losses only once during its 120-year long history (Thompson and Macklin, 2009). The stock exchange responded instantly. BHP share price fell below US$ 10. The company fell from its first place in the ranking of Australian companies to the fourth position.

The situation deteriorated to such an extent that rumors began circulating that other major companies began thinking of possible acquisition of BHP. Of course, not everyone could afford such luxury, but the companies such as Rio Tinto began to entertain these ideas. John Prescott tried to make substantial changes. He invited Bradford Mills to move from San Francisco to Melbourne to take the lead in formulating a new strategy for the company. In the high-level BHP management meetings, John observed that Brad was objective in describing the problems of the old limping vessel called "BHP" (Thompson and Macklin, 2009). In Australia, there was sharp criticism toward the leadership and their management style. The steel business was one of the stumbling "blocks" in the company's portfolio. This unit was underperforming and no longer generating significant profits. Despite the fact that the steel division had taken up much of Mr. Prescott's attention over the past 7 years, it became apparent that this unit should be sold off. In the face of prevailing market conditions, it was necessary to get rid of it, as well as from other inefficient businesses in the company. John made a proposal to the Company Board of Directors, chaired by Jerry Ellis at that time, to spin off this unit as a separate company. This idea was rejected by the Board and Mr. Prescott resigned as CEO in March 1998.

The BHP Board of Directors conducted a global search for candidates to the CEO position. The Chairman of the Board of Billiton, Brian Gilbertson, South African by origin, was among the candidates in consideration. Ultimately, however, the board appointed Paul Anderson, American, from Duke Energy as the head of the company. Paul previously enjoyed a reputation as a company savior specializing in recovering the firms from deep crises (P. Thompson and R. Macklin, 2010).

Without waiting for new reassignment, BHP began making significant changes. The causes of the dramatic situation in BHP were not only from the external problems such as falling mineral prices but also from the domestic issues, particularly in the area of the company management. Outside qualified observers suggested that the company turned a blind eye to "some of the decisions" which were made by some managers to further their personal ambitions in the company. The BHP Minerals Division, based in San Francisco, was criticized for isolating itself from the central management by making erroneous, unilateral decisions. The new policy was that the upper echelon of the company leadership would meet under one roof at the headquarters in Melbourne to avoid a similar situation.

For more than 100 years of its history, Australians drove "the big Australian." However, this time the leadership decided to depart from this custom. Now the main question was not about the leader's roots, but instead, how quickly the new leader could "pull" the company out of its quandaries.

In December 1998, Paul Anderson officially received his appointment to the post of President and Chief Executive Officer of BHP. Within 5 years of his leadership, he promised to double the value of the company and hence increase the value of its shares. Paul came to rescue of the company. With his leadership, the company "turned over a new page." After writing off almost 8 billion dollars, at that time an impressive amount even for a big company like BHP, the shareholders and the board of directors wanted to see a fundamentally new approach. Paul Anderson immediately got his "feet wet" and began a critical review of all company activities. His principal diagnosis was that the company had incurred losses due to inefficient production units. To remedy this situation, it was necessary to "stop the bleeding" by switching from value destructive to value creating operations. All activities of the company were subject to a critical review. Any initiative that did not directly lead to making a positive impact on the company's bottom line was subject to scrutiny.

Hugo Dummett headed the exploration division of BHP. The exploration also started conducting revisions for the approaches in their business. At that time, BHP enjoyed an excellent reputation in the mining exploration world due to a number of prominent discoveries made in the past such as Escondida, the Bowen coal basin, Ekati, etc. Many junior companies simply followed in the fairway of BHP staking claims near the ground controlled by BHP. An annual exploration budget for BHP in the second half of the 90s passed the milestone of US$ 100 million dollars. There were numerous active exploration programs in different parts of the world, but there was no precise system of selection criteria for the most successful projects. The newly designed policy was to ensure that the limited funds available for exploration would go to the most prospective discovery opportunities around the world.

The ultimate objective of the exploration unit was not merely conducting exploration but to deliver discoveries of new deposits of size and grade, which would constitute an unambiguous value to the company and its strategic development plans in the long run. The Geological Exploration Division was renamed simply to Discovery.

Second, the management determined the range of strategic mineral commodities, which it would pursue on the path to discovery. The list of strategic minerals included those that directly met the needs of the production units of the company: namely, copper, silver, iron, aluminum, nickel, diamonds, and coal. Discovery did not engage in oil exploration. Because each region advocated for their projects, the balance to get better and more objective decisions was to invite independent specialists on a variety of minerals which could help to provide an unbiased expert opinion. Thus, the structure of the Discovery Unit envisaged new positions for global managers in certain commodities.

Third, the Discovery stated that it targets only world-class deposits. All mineral deposits fell into several classes Tier 1, Tier 2, and Tier 3. It was necessary to determine their parameters. Interestingly, the definition of first-class Tier 1 deposits was about the cost of producing mines, built on such deposits. It defined that those positioned in the first quarter of the global list would rank as Tier 1, which meant that they would have to be profitable under all scenarios possible for mineral commodity price fluctuations on the world market. Deposits of the second class in the global price fluctuations could be productive only partially. Under low commodity prices, they would lose their profitability. It was estimated that in 1999 globally, on average, five discoveries of Tier 2 class deposits and 1.3 discovery of Tier 1 class deposits occurred every year (see Fig. 8.2). It is noteworthy that the exploration cost of deposit discoveries in the 90s significantly increased (Schodde, 2010).

In recent years, this picture even became more evident (R. Schodde, 2010 and 2012). The definition of Tier 1 deposits for various types of minerals somewhat differed, but the overall concept was based on the notion of the value of the deposits and their capacity to generate more than quarter of billion dollars after tax. The lifetime of such projects should be more than 15 years, and preferably more than 20 years. Subsequently, detailed definition of Tier 1 and Tier 2 deposits would be refined. In the meantime, for comparison, better understanding and interpretation, the reference to real typical examples of known Tier 1 deposits included Escondida and Chuquicamata in Chile, Grasberg in Indonesia, and Olympic Dam in Australia. For diamonds, it was kimberlite pipes Jwaneng and Orapa in Botswana, Udachnaya and Jubilee in Russia, and Ekati and Diavik in Canada. For sulfide nickel, it was October and Talnakh in Russia, Sudbury in Canada; while for iron, Mount Whaleback and

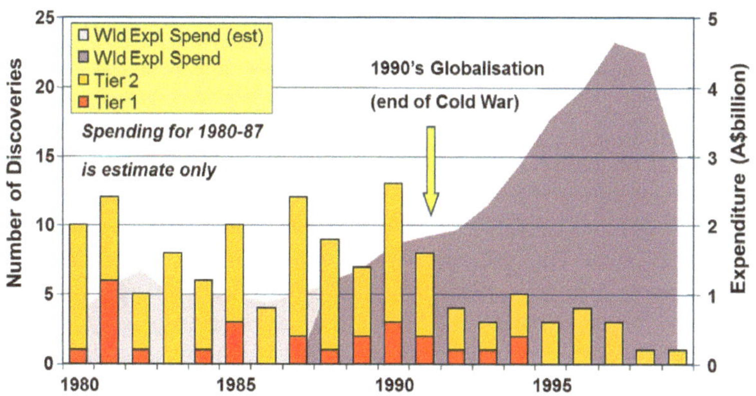

FIGURE 8.2 Statistics on discoveries of Tier 1 and Tier 2 deposits in the global mining industry.

Yandi in Australia. Examples of coking coal deposits were named Goonyella and Blackwater in Australia.

To maximize the best outcome from exploration, it is evident that the process required a system for selection of the best projects, which could consequently lead to the ultimate goal of discovering Tier 1 deposits. To do this efficiently, the leadership introduced a screening process based on a ranking system of exploration projects. The focus was on projects with Tier 1 potential. Third class projects were destined for closure. Ideally these projects would be sold but if there was no buyer then the property would be liquidated. The second-class projects' fate was dependent on the availability of investment funds. The company was ready to continue participating in the project but was not prepared to spend their own money on the continuation of exploration. In other words, for Tier 2 projects, we had to find investors who would be in a position to fund further exploration and share the exploration risk. In the absence of investors, the project was also subject to sale or closure, as in the instance of the third-class deposits.

The overall expectation was that new technological advancements in the exploration methods could facilitate BHP's quest for improvement and efficacy in their exploration efforts. "Falcon", which was mainly a gravity-based geophysical method, was to become such a technology. Gravimeters were to be placed on airplanes or helicopters. Effect of substantial increase in the survey performance would allow covering big areas in a short stretch of time. Sander Geophysics developed the technique in 1997 and BHP had exclusive rights to use it during a certain period. It was perceived to be a competitive edge for the Discovery unit of BHP, which could potentially have a unique opportunity through Falcon to see what the others could not do. A number of the strategic minerals included iron, coal, diamonds, nickel, or the types of mineral deposits, where the ore bodies had a noticeable contrast with the density of the surrounding host rocks, it meant that Falcon could quickly delineate the anomalous areas for subsequent drill testing. Hence, the intensive use of this technology could thus decrease the chain of steps in the exploration process.

We should note here that in the history of the exploration, there were many examples of magical philosopher's stone on how to successfully search for and predict mineral deposits, allowing explorers "seeing" or "feeling" the signal from the depths. During the time of Georgius Agricola in the 1500s, it was thin willow twigs and in recent times, there were attempts to generate attractive projects by intensive data processing. It is a so-called phenomenon of "silver bullet"—some almost magic extremely efficient tool to see through the rocks of the earth crust. Yes, data processing was necessary. However, it could not completely replace exploration fieldwork. Innovation data processing, particularly to consolidate view and to understand geology in three-dimensional space, was an essential element. Applying new efficient methods of exploration improves chances for successful discovery. Even though the general trend is that discovery costs are increasing and the rate of major discoveries is decreasing, with the right approach important discoveries, will continue to be made. This was the message that the Discovery group presented to the BHP management. It is noteworthy that the exploration cost of deposit discoveries increased significantly in the 1990s (see Fig. 8.2).

Tier 1 deposits do not always become immediately apparent at the onset. In the modern era, the mineral discovery process often begins with a geological concept which is then followed up with prospecting, or geological reconnaissance, and then a comprehensive exploration program. As a rule, discovery begins with the success of detecting mineralization, which on further investigation evolves into the third-class and/or second-class ranks and then subsequently becoming the first-class Tier 1 discoveries. The system of elimination and selection of the real candidates for first-class discoveries is a critical component of the exploration process. The question of what projects deserve further funding and what projects should be suspended or cut is highly sensitive because the correct answer requires foresight, acumen and making the decision requires courage.

To make this selection process less subjective, a systematic approach for assessing each stage of advancement along the exploration pipeline was developed. Each successive stage of the pipeline was punctuated by a filter or "key decision point". To pass to the next stage, a prospect had to satisfy pre-determined criteria. A facsimile of this model was taught long ago in the Soviet geological schools to minimize inappropriate use of exploration funds. This discipline facilitates the selection of projects that merit further financial support and identifies those projects that are unlikely to become Tier 1 discoveries. Funding should be stopped for the latter category. The exploration programs have to be organized so that these decisions can be made relatively early to minimize the unnecessary costs of exploration work.

Arguably more complex than differentiating prospects of the same commodity with Tier 1 potential from those with only Tier 2 and Tier 3 potential, is the comparison of Tier 1 potential projects at different exploration stages and with different types of mineralogy. For example, copper porphyry and diamond projects are fundamentally different from each other. It was, therefore, necessary to develop a common database so that one could compare various projects.

Fourth, the leadership introduced the risk management system. The list of risks surrounding any mineral deposit includes political, environmental, social, and technical risks, among others. Previously, geologists used technical parameters as the main criteria. The initial assumption was that, in case of a successful discovery, the company would be able to manage the other risks associated with the project as it progressed toward production. However, in reality, especially in the modern days, it was becoming obvious that ignoring such risks at an early stage could result in some advanced projects falling into a state of paralysis due to underestimated risks and project challenges. Effective management required identification of the risks and development of the strategies for their management at a very early stage to preempt or minimize problems due to unforeseen circumstances. To better manage risks, BHP created a team within the Discovery unit with the task of formulating the strategy; hence, the Discovery Strategy Group was formed.

It is well known in the mining industry that exploration geologists tend to be persevering in their quest in making a discovery. Many also tend to form a strong attachment to their projects and can, in few cases, be guilty of recommending further exploration although results indicate that the project is essentially "dead" and that it is beyond "resuscitation".

The Discovery Group intended to not only make its own in-house discoveries but also to monitor the competitive field primarily for junior and midsize companies to identify projects with the world-class discovery potential in their portfolios. To address this, the exploration Discovery Group introduced a commercial function, which was supposed to keep track of what happens in the competition space and, in case of detection of promising projects, to conclude profitable business deals. Analysis of the state of the mining industry showed that not only major mining companies but also some small, so-called junior companies, delivered the first-class discoveries. The smaller companies, due to their higher mobility and higher tolerance to risk, and the fact that they are not restricted to focusing on properties that have the earmarks of a world-class deposit, can have the luxury of persevering on smaller deposits that could turn out to be large enough to become of interest to a major company. It was, therefore, necessary to build strategic alliances with junior companies that demonstrated their ability to achieve exploration success on the discovery path. Such an approach allowed Discovery to rely not only on their intellectual potential but also utilize usually inaccessible potential of the other companies.

BHP Discovery was also subjected to structural changes. Instead of the costly and inefficient regional system, the concept of newly consolidated regions came into effect. Global exploration was to be conducted out of three, already existing hub offices. The office in Vancouver was responsible for North America/Europe region, which also included Northern Asia. The office in Brisbane was the center of the region of Australia, Africa, and Asia; Central and South America were controlled from the Santiago office. Interestingly, the separation line between the North and South Asia took place just along the State border between Russia, Kazakhstan, along with other Central Asian and Caucasian republics subdividing them into in the North Asian region, and China, Mongolia, Pakistan, Iran, and Turkey, into the South Asian region. Under this new arrangement, the Mongolian projects went under the umbrella of the Australia-Africa-Asian region controlled by the Brisbane Office.

Discovery tried to demonstrate to the new leadership of BHP that the changes made by the division's procedures in conducting geological exploration would result in a greater frequency of making Tier 1 discoveries in the future. To confirm its commitment, Discovery pledged to deliver two first-class discoveries every 10 years. This was a noble commitment and one that would be very challenging to deliver.

In May 1999, the company announced its decision to suspend North American exploration programs. The company decided to close its office in San Francisco. Financing of BHP global exploration programs decreased by more than 50%, which reduced the annual budget to about US$ 50 million dollars. Compared with other major competitors, such as Anglo American with a budget of US$ 136 million dollars in 2001, it was a significant reduction. The management introduced a new approach for the allocation of funds to the BHP Discovery Division. One portion of the budget supported the operational costs of the exploration offices and existing projects, and the other portion was to fund new proposals from the so-called Opportunity Fund. This part of the budget was open to any program that could qualify for financing from the Fund on a competitive basis. Financing was granted on a competitive basis to the projects that could demonstrate potential for Tier 1 discovery.

Paul Anderson after taking over the reigns of CEO announced the sale of the company's underperforming assets valued at US$ 2 billion dollars. In June, the company also recognized that not all its efforts to solve the environmental problems at the Ok Tedi mine in Indonesia were successful and the company was considering withdrawal from the project. The Escondida mine in Chile announced a discovery 5 km north of the main mine named Escondida Norte with a resource of 676 million tonnes at average grade 1.03% copper. About 15% of the resource was amenable to heap leaching. Also, management decided to increase copper production at Tintaya in Peru by commissioning oxide ore that was supposed to increase copper production bringing it to the level of 110,000 metric tonnes of copper per year.

BHP now gave preference to the exploration projects near its operating mines and started reducing its exposure to projects with high degrees of various types of risk, whether technical or environmental. Management required the exploration units to articulate their goals and objectives for projects, as well as the process of achieving them. After the closure of Hartley and the decision to withdraw from the Ok Tedi mine, the company expressed its preference to consider only exploration projects in geographic locations with controlled risks.

After critical revision of all exploration programs from a new perspective, many of them had to either be reduced or closed. All exploration work in the Russian Federation was terminated. The largest of them was the search for copper and gold deposits in the Magadan region on the Upper Uptar copper porphyry deposit, on the gold-bearing areas of Maliy At-Uriah and Vodorazdelny, and on the Glukhariniy gold deposit. After it became clear that the BHP ceases funding for these projects, the company began searching for potential investors, primarily from the junior sector.

The sharp edge of exploration budget cuts also affected the Mongolian program. The Discovery management team categorized the Mongolian project

Oyu Tolgoi as a second-class porphyry deposit, and, thus, BHP was seeking investments from the outside to further fund the exploration program there.

Dramatic changes took place across the company. The BHP Board of Directors also saw some changes. Jerry Ellis stepped down as Chairman of the Board of Directors and Donald Argus succeeded him in this position (Thompson and Macklin, 2009).

Chapter 9

Search for Investors

Douglas J. Kirwin *was born in Australia, received bachelor's degree in mining geology from the Queensland University and master's degree from James Cook University in Townsville, Queensland. Douglas worked as Managing Director of the International Geological Services Pty. Ltd. for several years. He also worked at Anglo American and Amax. Douglas was engaged in the assessment of ore fields in America, Asia, and Europe and managed several projects in Southeast Asia and Australia. In September 1995, Douglas became the head of exploration of Ivanhoe Mines.*

Once it became clear that to continue exploration work at Oyu Tolgoi it was necessary to find investors; an intensive search for companies wishing to join BHP was launched. Initially, our attention was drawn to significant copper producers such as Rio Tinto, Phelps Dodge, Western Mining, Minorco (i.e., Anglo American), Freeport, and ASARCO. We also considered some junior companies engaged in active exploration in Mongolia, such as Troy Resources, Resolute, Cascadia, and others; however, we did not cherish serious hopes on them.

The plan was that in the case of finding an interested party or investors we could carry out additional verification on the sites that had not been thoroughly explored, focusing on the areas where the chalcocite blanket was expected to be at shallow depths, based on BHP's previous drilling. A reverse circulation drilling could quickly determine whether there was still significant potential for additional secondary enrichment at the project. Indications from previous

Discovery of Oyu Tolgoi. https://doi.org/10.1016/B978-0-12-816089-3.00009-3

drilling were that there was potential for additional hypogene copper-gold mineralization. The project geologists suggested paying attention not only to surrounding areas directly near Oyu Tolgoi, but also to all the anomalous sites within the license area. The project geologists were optimistic that execution of such a drilling program could upgrade Oyu Tolgoi to a first-class deposit. The conviction among the geologists working on the project was that the real potential of the deposits had not been assessed adequately yet, and more pleasant surprises were waiting for the persistent explorer.

The proposed exploration program was to conduct basic exploration methods such as detailed magnetic and gravity surveys, combination of rock and soil geochemistry and additional IP geophysics. Previously delineated anomalous areas for chalcocite mineralization would be drilled by a series of shallow drill holes through the cover to adequately map out the top of the bedrock under Cretaceous postmineral cover. Based on the favorable results, it was proposed to conduct a more focused reverse circulation drilling program to delineate the chalcocite blankets under the cover rocks.

The exploration team believed that with this program, the potential chalcocite enrichment zones could finally be adequately assessed. The investors also had an opportunity to delineate further the extensions of the hypogene mineralization on the property.

In June 1999, the delegation of Minorco visited Oyu Tolgoi deposit. A consultant for Minorco was Dick Sillitoe, a renowned expert on copper porphyry deposits. After visiting Oyu Tolgoi, Dick praised the quality of the work carried out by the BHP project team, noting that for the money BHP spent at the project, it managed to complete much more work than any junior company he had a chance to visit before. It was a very valuable compliment for the Mongolian team of BHP led by Don Schissel, Sergei Diakov, and Barrie Bolton.

Based on the results of its activities in Mongolia during 1998/1999, the Mineral Resource Authority of Mongolia (MRAM) acknowledged BHP Minerals Branch as the most successful company for investment and performance in the field of mineral resources in Mongolia. In this regard, the company received the official diploma certificate from the Agency of Mongolia's mineral resources (Fig. 9.1).

Following the reduction of BHP's funding in Mongolia, Barrie Bolton returned to Australia. After his departure, the reins of BHP Minerals Office in Ulaanbaatar went to S. Sanjdorj. On February 17, 2000, the licenses issued for Oyu Tolgoi (license No. 210 area 135,987 ha) and Ih Shanh (license No. 211 square 44,809 ha) reached their 3-year term. The BHP Branch submitted a request for an extension to MRAM until February 17, 2002, which was subsequently granted. The outline of the licensed areas remained unchanged.

In search of investors, the management decided to present the project at the annual Prospector and Developers Association of Canada (PDAC) Conference in Toronto. Incidentally, the Conference organizers, one of whom was Rod

FIGURE 9.1 BHP Best Investor of the Year 1998/1999. Left: the Certificate of MRAM; right: D. Enkhbold, representative of MRAM, Douglas McGay, Chairman of Investors Union in Mongolia and Barrie Bolton.

Thomas, who previously worked for BHP, invited the company to exhibit drill core from Tolgoi Oyu.

The PDAC conference is the most significant exploration forum in the world and hosts prospectors and investors from all over the globe. Attendances can be as high as 30,000 delegates. The “core shack” is one of the most popular venues at the PDAC convention. Here, for a nominal fee and with approval from the PDAC Committee, drill core, and other promotional materials such as geological maps, and posters describing projects from all around the world can be put on display. In most cases, advanced exploration projects are presented, usually with the objective of attracting investors to fund those projects.

The PDAC Committee approved the Oyu Tolgoi presentation materials, and BHP was given 1 day to present the project to the world. It is estimated that 50–100 delegates visited the kiosk (Fig. 9.2), many of whom expressed a keen interest on what was displayed. Most of the visitors were representatives from mining and exploration companies ranging from juniors to some of the world’s largest companies. This was one of the first years that several companies with projects in Mongolia were present. At that time, limited information was available about the investment climate in Mongolia. Junior companies, including Cascadia Mining Inc., AGR, Java Gold Corp., who also had projects in Mongolia, visited the kiosk.

Douglas Kirwin, Senior Vice President of Ivanhoe Mines, who led the Exploration Division of the company during the Conference, stopped at BHP kiosk and expressed a profound interest in information about Oyu Tolgoi. Fortuitously, Ivanhoe was looking for new opportunities in both Mongolia and China during those years.

FIGURE 9.2 BHP kiosk at PDAC Conference 1999 with the drill core from Oyu Tolgoi deposit.

At that time, Ivanhoe Mines led the development of the Monywa copper porphyry deposit in Burma, now Myanmar. It was a joint venture with Burma's military government, which caused an adverse reaction in the public. Search for alternative copper porphyry deposits in a more appropriate jurisdiction was a priority for Douglas.

The zone of secondary sulfide enrichment at Central Oyu Tolgoi was of great interest to Douglas. Being much smaller than BHP, Ivanhoe's target size thresholds were comparatively lower. Conceivably, a discovery of several larger chalcocite blankets could constitute a project of interest to the specified parameters of Ivanhoe Mines.

A number of other companies expressed an interest in Oyu Tolgoi. In May 1999, we prepared comprehensive data packages and circulated them to all potential candidates for their review. The package also included data from the Ih Shanh prospect and a series of copper and gold mining licenses at Khubsugul, Zamtiin Khad, Delgerekh, Berkh, and Khentei, which BHP was holding at that time. Some of the large companies, such as Minorco, Phelps Dodge, and, mainly, Western Mining expressed a great deal of interest. On the other hand, Rio Tinto seemed to be primarily interested in gathering information and did not have serious interest in the project at that time. Some of the interested parties even visited Oyu Tolgoi.

For BHP, Western Mining was an exceptionally suitable option. It was also an Australian company with a similar business culture and with analogous management approaches to develop mineral deposits. Of course, Western Mining was much smaller compared with the "Big Australian", but regarding their technical level, it had a brilliant reputation both within Australia and internationally for being geologically shrewd and intellectual with numerous of successful discoveries under their belt. Western Mining Company had several nonferrous mines, including the Olympic Dam deposit, which was one of the most massive copper-gold-uranium deposits in the world, discovered by the company geologists. The successful discovery list of Western Mining included a copper porphyry deposit Tampakan in the Philippines and a number of nickel deposits in Western Australia.

After visiting Oyu Tolgoi, the Western Mining technical team had a very positive opinion of the deposit and recommended to their management a proposal to enter earn-in to the project via an option. As the proposal moved through the hierarchy of the company, it suddenly stalled at the company Board of Directors' level, and at the desk of the CEO Hugh M. Morgan. The leadership of the company decided that they preferred to direct their investments in the countries with the established investment climate and controlled risks. Instead, Western Mining took an extensive land package in the Canadian province of Quebec focusing on nickel exploration.

Ironically, BHP Billiton prompted by the desire to control the giant Olympic Dam deposit acquired Western Mining in 2005. Following this acquisition, all projects from Western Mining transferred to BHP Billiton. Had the 1999 deal been consummated, then Oyu Tolgoi would have been one of them.

At this time, the terms of an agreement were still open to discussion and BHP was prepared to offer the investor various flexible payment options, including the possibility of postponed option payments. BHP would be prepared to take back the project on the expiry of any mutually established deadlines. The seller also guaranteed the acquirer the transfer of the skilled local personnel, together with the project so that the "new boss" could minimize the difficulties of the transitional period. The local BHP team proved that it was technically competent and knowledgeable in all aspects of the project and they would be essential for the future success of the projects.

Investors were required to provide a program plan designed to confirm the presence of mineralization on the exploration licenses in conjunction with time and expenditure commitments to fulfill the program. BHP committed to provide its entire Mongolian database and any machinery and equipment related to its projects in the country. If necessary, BHP would offer investors a guarantee to visit the BHP projects independently should they want to verify the facts. In addition, BHP committed to provide all available database of earlier work conducted in Mongolia, including equipment within the framework of company projects in the country.

The deadline for responses was set for September 11, 1999. Ivanhoe Mines responded positively and suggested they were open to discuss the details and conditions of the agreement. The first direct contact was Edward Rochette, then Senior Vice President Ivanhoe Mines for legal matters and administration. In October 1999, a meeting took place in Melbourne, where details of the future agreement were the subject of discussion. After multiple interactions, Ivanhoe Mines prepared a proposal to BHP.

Chapter 10

Agreement With Ivanhoe Mines

Robert Friedland *was born in Chicago. In 1974, he graduated from Reed College in Oregon. While in college, he became friends with Steve Jobs, the founder of Apple Inc. On weekends, Robert with Steve spent time on his uncle's farm picking apples; hence, the name of Jobs' company Apple Inc. Robert was the President for Galactic Resource, which led the development of the Summitville mine in Colorado. In 1990, a cyanide leak occurred, resulting in mine closure and a court case. One of the most prominent pages in Robert's career was the discovery of a large nickel deposit in Labrador Voisey's Bay and a very successful sale of it to the Canadian nickel company Inco Ltd. In 1994, he founded Ivanhoe Mines Ltd and registered it on the Toronto stock exchange. Robert's strength is the ability to raise funding for geological exploration.*

Discovery of Oyu Tolgoi. https://doi.org/10.1016/B978-0-12-816089-3.00010-X

Charlie Forster *was a graduate of the University of British Columbia in Vancouver, Canada. He received his bachelor's degree in geology in 1970. He was employed by the Texas Gulf Corporation and Echo Bay. In China, Charlie worked as manager of exploration for the Yunnan Mining Limited and Diamond Works companies. For Ivanhoe Mines, he occupied a position of Senior Vice President in charge of the Oyu Tolgoi project. He was an authorized specialist in Canadian national requirements 43–101. He participated in several critical decisions on behalf of Ivanhoe that had effects on the future fate of Oyu Tolgoi. At the time of the writing of this book, he was the Chief Geologist and Exploration Manager of ENRC in Africa.*

On May 8, 2000, Ivanhoe Mines issued a statement that it had signed an agreement with the Broken Hill Proprietary Limited (BHP) company, which provided Ivanhoe Mines with an opportunity to acquire a 100% stake in the Oyu Tolgoi copper deposit in Mongolia's South Gobi. Under the terms of the agreement, Ivanhoe Mines was obligated to spend US$ 6 million dollars on exploration activities over a 7-year period and pay a sum of US$ 5 million dollars if BHP did not come back to the project. On fulfilling these requirements and with the approval of the Government of Mongolia, the 1,360 square kilometer Oyu Tolgoi license area Oyu would be transferred to Ivanhoe Mines. Ivanhoe's statement noted that BHP spent about US$ 2 million dollars conducting geophysical, geochemical, and mapping and diamond drilling programs.

Ivanhoe believed that Oyu Tolgoi had good potential for increasing the volumes of both secondary copper and primary copper-gold mineralization. To confirm this, it planned to conduct a program of 4,000 m of reverse circulation drilling. The company believed that the project had the potential to identify sufficient copper mineralization for open-pit development and that the supergene ore was amenable for low-cost copper production by electrowinning, which was the process used by Ivanhoe's 50% owned Monywa project in Burma. Ivanhoe also expected that they could increase profitability by recovering gold in the leached porphyry by the heap leaching method.

Ivanhoe was obligated to pay US$ 1 million dollars on the signing date of the option on May 8, 2000, and they were obligated to pay the remaining US$ 4 million dollars on the first anniversary of the option agreement. During the first phase of the agreement, from 2000 to 2003, Ivanhoe was planning both detailed diamond core and noncore reverse circulation drilling programs within the North, Central, and South Oyu Tolgoi areas. Ivanhoe Mines estimated that the minimum cost during the first year would be US$ 500,000 dollars, which would then increase to US$ 1 million dollars in the second year and to US$ 1.5 million dollars during the third year.

During the second phase of the agreement, from 2003 to 2007, Ivanhoe was obligated to spend US$ 3 million dollars on exploration activity, of which US$ 1.5 million dollars had to be spent on exploration outside the Central Oyu Tolgoi chalcocite zone. If at any time during the second phase of the agreement both parties agreed that 250 million tonnes of mineralization grading 1% copper or more had been delineated, which could be processed by a simple electro winning process, then BHP could buy back 40% in the project. If the detected mineralization was 300 million tonnes of 1% copper or more requiring a conventional copper extraction through enrichment, then BHP would have the right to buy back 60% of the stake. In both cases, BHP would be obligated to pay Ivanhoe a sum 3 times greater than Ivanhoe Mines' expenditure on exploration work during the first and second phases of exploration. BHP would also be required to refund all payments received from Ivanhoe at the time of the project purchase. After that, BHP and Ivanhoe Mines would be required to set up a joint venture to continue with the project. If BHP did not exercise its right to buy back into the project within a period of 60 days from the date of the announcement, BHP would maintain the royalty rights of 2% of the future gross production from the project.

During the whole earn-in period, Ivanhoe was obligated to maintain the license and to comply with all requirements for payment of the license fee as well as to provide the required reporting throughout the duration of the works. If during the first term Ivanhoe had to spend more money than determined by agreement, then the overrun was to be accounted against the cost commitment for the second term. In case of default, Ivanhoe would lose all rights and had to abandon the project completely.

Ivanhoe was encouraged to use BHP's local staff on the project. The terms of the contract were not to be inferior to those that BHP had provided for its local staff. BHP Minerals also proposed that Ivanhoe purchase all its equipment and materials in Mongolia at market price.

Robert Friedland, Chairman of Ivanhoe Mines, supervised the negotiations, which were ongoing from October 1999 to May 2000, between his team and BHP from behind the scenes. Marcus Randolph, a Senior VP of BHP, was providing senior management support on behalf of his company. Despite the fact that Robert and Marcus had been on opposite sides of a lawsuit, in the past, they maintained a professional relationship during the negotiations.

Charlie Forster, an outstanding Canadian geologist, with a great enthusiasm for mineral exploration and discoveries, was given the responsibility of Oyu Tolgoi project supervisor.

Under the terms of the contract, Ivanhoe Mines kept BHP's local Mongolian team on their project staff list. That was an important condition and guarantee for further successful advances of the project. Therefore, the exploration team of D. Garamjav, T. Munkhbat, and G. Sarangerel, led by S. Sanjdorj, transferred to Charlie's team and became employees of Ivanhoe Mines. Subsequently, this played a significant role in the future of Ivanhoe Mines at Oyu Tolgoi project. The project would face new challenges and dramatic moments in its new history under the banners of Ivanhoe Mines.

Chapter 11

First Period of Ivanhoe Mines Investment

After signing the agreement with BHP, Ivanhoe Mines immediately began organizing their field camp and started preparation for exploration. Drilling already began in June 2000 and continued until September. During this time, Ivanhoe drilled a total of 8,828 m in 109 reverse circulation drill holes. Drillhole OTRC105 intercepted additional chalcocite mineralization on the flank of South Oyu. In general, according to Ivanhoe, the results were encouraging. New drilling information suggested that the volume of chalcocite mineralization could almost double compared to BHP's previous calculation.

Under the terms of the Agreement, Ivanhoe regularly submitted project reports to BHP. Periodically the BHP representatives visited the project site for monitoring and reconciliation of the reported results with the field activities. The Oyu Tolgoi exploration license was still in the hands of BHP, and therefore all responsibilities for reporting the exploration activities and their impact on the local environment remained with BHP. The time for payment of the annual license fee was approaching. As indicated above, under the Mineral Law at the fourth and the fifth anniversary of the license the fee doubled, rising from US$0.5 per hectare to US$1.0 per hectare and subsequently increasing to US$1.5 dollars during the last two annual terms. Because the size of the Oyu Tolgoi license area remained at 135,987 ha, it was time to consider the possibility of reducing the size of the license area and consequently the fee payments. BHP and Ivanhoe were able to agree on a significant reduction of the original license area. BHP's application to Mongolia's Mineral Resources Agency (MRAM) to reduce the license was accepted. The map on Fig. 11.1 shows the outlines of the areas that were retained for further exploration. Table 11.1 summarizes the coordinates and dimensions of the reduced areas as submitted to the Government. All terms and conditions of the original license of Oyu Tolgoi went to the license areas within the newly proposed contours.

It should be noted that there seemed to be a degree of disregard for exercising due diligence in selecting the ground to be excised from the Oyu Tolgoi license. It was known that parts of the "condemned ground" were

Discovery of Oyu Tolgoi. https://doi.org/10.1016/B978-0-12-816089-3.00011-1

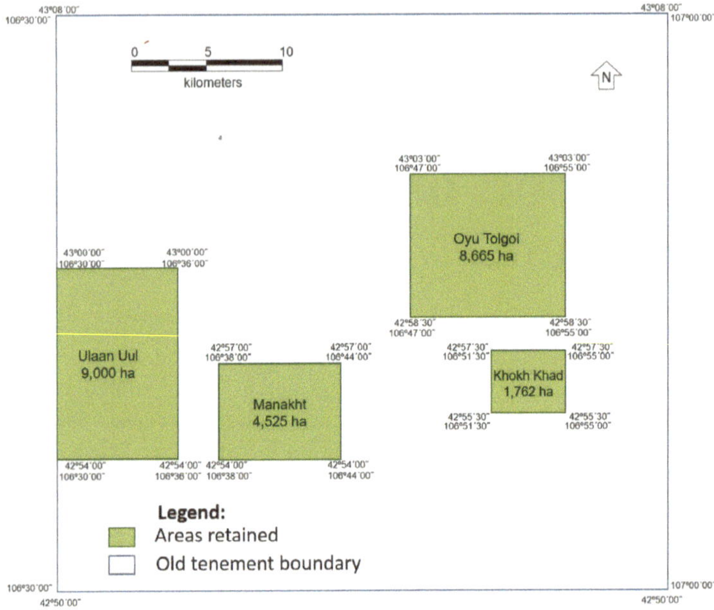

FIGURE 11.1 Map showing the outline of Oyu Tolgoi license and the retained areas.

inadequately explored to rule out the possibility of significant mineralization under cover. A more sensible approach, at this stage of exploration, would have been to only surrender the parcels of land that were known to be underlain by granitoid massifs. The granitoid complexes in the Oyu Tolgoi region are posthydrothermal mineralizing events and have no possibility of hosting the targeted deposit type. By doing so, the licensed areas underlain by volcanics, including those covered with younger Quaternary rocks, would have remained under Ivanhoe's control for further investigation until it became clear in which direction the mineralization extended.

In 2000, Hugo Dummett, VP of Exploration for BHP, was replaced by Tom Whiting. In 2001, Hugo Dummett became Deputy Chairman and Executive Vice President of African Minerals, a subsidiary company of Ivanhoe Capital Corporation. It was a boost for Ivanhoe that Hugo could continue to have influence over the Oyu Tolgoi project. He was also elected as President of the Society of Economic Geologists (SEG). Tragically, following his next visit to Oyu Tolgoi in August 2002, Hugo died in a car accident while returning to his base in South Africa. The World Bank forum in Johannesburg was taking place at the time of his arrival to South Africa. The circumstances were that the hotels in Johannesburg were full and despite being exhausted and jetlagged after a long flight from Mongolia, Hugo decided to drive to his sister's place. During this drive, his vehicle left the road and sadly the great explorer of modern times Hugo Dummett was killed.

TABLE 11.1 The Coordinates of the Retained License Areas of Oyu Tolgoi

	New Coordinates							
	Easting			Northing				
Prospect	Grad.	Min.	Sec.	Grad.	Min.	Sec.	Initial Area (ha)	New Area (ha)
Oyu Tolgoi	106	47	00	42	58	30	135,987	8,665
	106	47	00	43	03	00		
	106	55	00	43	03	00		
	106	55	00	42	58	30		
Khokh Khad	106	51	30	42	55	30		1,762
	106	51	30	42	57	30		
	106	55	00	42	57	30		
	106	55	00	42	55	30		
Manakht	106	38	00	42	54	00		4,525
	106	38	00	42	57	00		
	106	44	00	42	57	00		
	106	44	00	42	54	00		
Ulaan Uul	106	30	00	42	54	00		9,000
	106	30	00	43	00	00		
	106	36	00	43	00	00		
	106	36	00	42	54	00		
Total							135,987	23,952

The drilling at Oyu Tolgoi continued. In addition, to the reverse circulation drilling, Ivanhoe also drilled three diamond core drillholes, OTD149, 150, and 159, respectively, within South, South West, and Central Oyu Tolgoi. Diamond drilling was necessary to validate the zones of hypogene mineralization with elevated copper and gold grades at depth that had first been discovered during BHP's first phase of drilling in 1997.

Drillhole OTD149 intersected 63 m of oxidized mineralization with an average content of 1.08% copper and 0.31 g/t gold. This was followed by a 71.5 m interval of bornite mineralization with 1.36% copper content and 0.18 g/t gold. Hole OTD150 was drilled across BHP's drillhole OT10 and intercepted 70 m of chalcopyrite mineralization extending to 508 m along the axis of the borehole with an average copper content of 0.81% and gold content of 1.17 g/t. OTD159, drilled in Central Oyu Tolgoi, essentially confirmed the earlier results. At a depth of 47 m, it intercepted a layer of chalcocite mineralization 49 m thick with an average content of 1.17% copper and 0.21 g/t gold. Interestingly, under the chalcocite blanket, there was a zone of hypogene copper mineralization containing covellite with a 252 m interval, which returned grades of 0.61% copper and 0.11 g/t gold.

These three important intercepts provided sufficient encouragement for Ivanhoe Mines to continue exploration drilling at Oyu Tolgoi. The primary mineralization in drillhole OTD150 was particularly inspiring and Ivanhoe's focus began to shift to South West Oyu Tolgoi. Here, dense pattern of holes were drilled to a depth of 500 m or more.

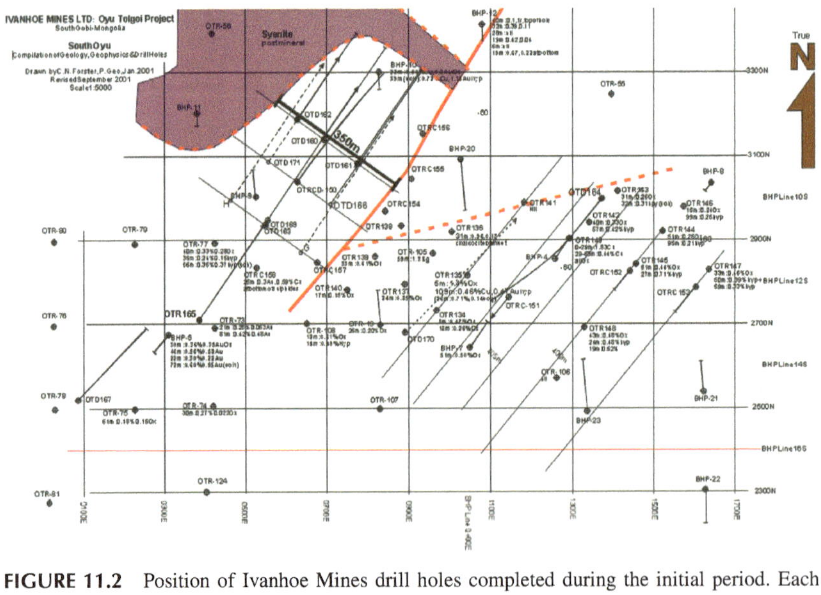

FIGURE 11.2 Position of Ivanhoe Mines drill holes completed during the initial period. Each grid cell is 200 × 200 m.

These new encouraging drilling results (Fig. 11.2) and the fact that a local Mongolian company now occupied the northern part of the former BHP-licensed area compelled Ivanhoe to reconsider their ground position. Ivanhoe decided to thoroughly explore the Devonian volcanic belts in the area still under its control, to the south and to east of the drilling.

In February 2002, the Ulaan Uul, Manakht, the Khokh Khad, and the remainder of the Oyu Tolgoi exploration licenses were approaching their second anniversary. It was necessary to renew the licenses for their final 2-year exploration extension as stipulated by the legislation. At that time, there was little doubt regarding new ground selection or reduction. Although Ivanhoe Mines lost some potentially prospective ground, particularly at the northern

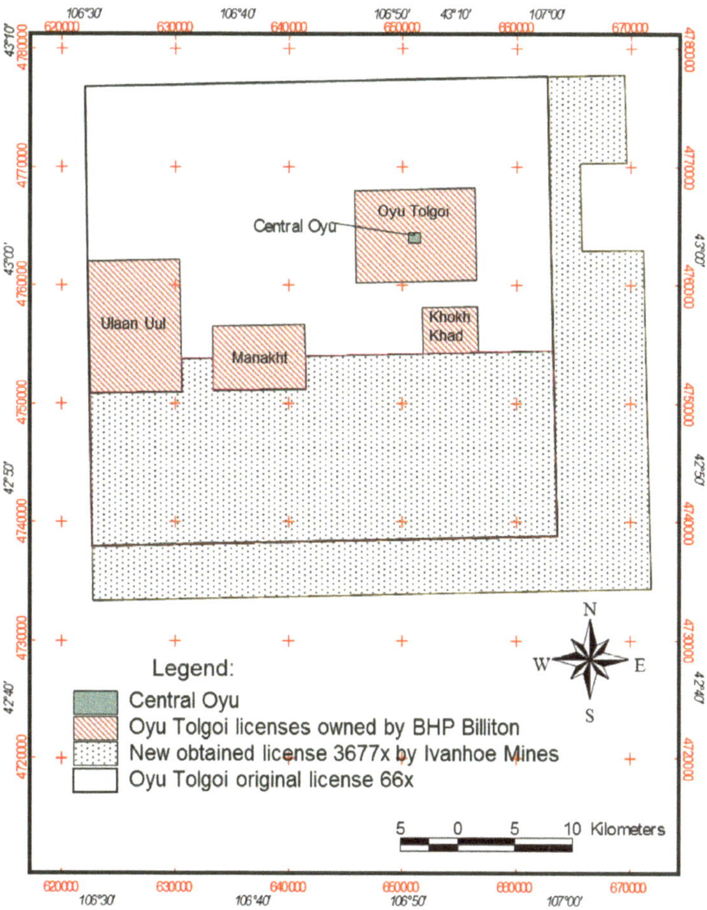

FIGURE 11.3 Limits of the license areas of Oyu Tolgoi in 2002.

and southern ends of the original license, they felt that still they continued to control the most interesting and prospective ground (Fig. 11.3). After making a commercial deal with the local Mongolian company, Canadian junior company Entrée Gold came in to take a neighboring land position. Later on, Ivanhoe Mines also concluded a deal with Entrée Gold.

Chapter 12

New Corporate Changes in BHP

Brian Gilbertson *received his Bachelor's degree in mathematics and physics at the University of Rhodes in 1963. In 1969, he received Master's degree in physics. In 1973, he obtained Master's degree in business from the University of South Africa. In 1992, he participated in the reorganization of the South African company Glencore and then in 1994, in the merger with the mining subdivision of the Danish company Shell Billiton. In 1997, Glencore sold its unit of nonferrous metals to company Billiton PLC listed on the London Stock Exchange, and in July 2001, together with Paul Anderson, Brian Gilbertson led to the merger of Billiton PLC with BHP Limited in Australia to form BHP Billiton PLC joint company.*

Charles (Chip) W. Goodyear *was born in Hartford, Connecticut, United States. He graduated from Yale University in 1980 with Bachelor's degree in geology and from Wharton Financial School in Pennsylvania in 1983. From 1983 to 1985, he worked for the Peabody coal company, where he rose to the position of Vice President of Finance. From 1983 to 1997, he was Vice President for financial affairs for the company Freeport-McMoRan mining company. From 1997 to 1999, he was President of the Goodyear Capital Corporation. Chip joined BHP in 1999 and until 2001 he was the Chief Financial Manager, from 2001 to 2003 the Chief Development Manager, and from 2003 onward as the Chief Executive Officer of BHP Billiton.*

Discovery of Oyu Tolgoi. https://doi.org/10.1016/B978-0-12-816089-3.00012-3

In 2001, BHP started a new cycle of their changes and transformations. In March, the sensational announcement was made that an agreement had been reached between two of the world's giant mining companies, BHP and the South African company of Billiton to form a merger. The merger would make BHP Billiton the world's leading diversified mining company with capitalization of US$38 billion dollars. Only 3 months later, the CEO of BHP announced that the basic integration of the two companies was complete. A new company structure came into effect and BHP Billiton assigned new managers to all its main posts.

The global marketing group was set up in two centers, one in The Hague and the other in Singapore. Of course, full integration process required more time to settle all of the problems from top to bottom. The primary task was to avoid duplication and to take overall advantage of each company's strengths, which organically were mutually complementary. Completion of the merger required spinning off BHP's steel units.

In the new structure, the strategy was that all units work toward improving their productivity. Initially, the main production units of the joint company were as follows:

- Aluminum (aluminum and aluminate)
- Nonferrous metals (copper, silver, zinc, and lead)
- Materials for steel production (coking coal, iron, manganese)
- Materials for production of stainless steel (chromium and nickel)
- Thermal coal
- Oil (oil, natural and liquefied gas)
- Steel

The Management Committee of the joint company was Paul Anderson, the former CEO of BHP, and Brian Gilbertson, the former CEO of Billiton. The company was registered on the London stock exchange as BHP Billiton PLC and the Australia stock exchange as BHP Billiton Limited. The company's headquarters was located in Melbourne.

The challenge for the Discovery Unit was to integrate two exploration groups with entirely different structures. The BHP group had a long history of in-house project generation and a reputation for the ability of making world-class discoveries. Discovery also supplied BHP's mining division through the acquisition of advanced exploration projects or mining companies. Billiton's approach to geological exploration was much more straightforward. Their elected formula was to liaise with junior companies, registered mainly on the Toronto stock exchange (TSX). Billiton also had a small geological research staff in the company, which was mostly a commercial unit engaged in the management of the alliance partnerships, as well as a small copper exploration group left from its recent acquisition of Rio Algom. Therefore, these two exploration groups seemed to complement each other in their approaches to exploration; there was not a lot of duplication between them. Therefore, the amalgamation of the two exploration groups into one turned out to be not that much difficult as initially perceived. As in the previous case with BHP, only first-class Tier 1 discoveries were significant enough for the joint exploration unit although the size of the requirements were even higher. With the merger, the spectrum of the minerals on the strategic list expanded.

In May 2002, Paul Anderson announced that he fulfilled his promises to the BHP Board of Directors and reminded the Board about his plans of resignation. The value of BHP shares since he became Chief Executive Officer did not increase significantly, but the merger of BHP and Billiton led to a sharp increase in capitalization of the combined company. The reins of the integrated company were intended to pass into the hands of Paul's logical successor, former head of Billiton and the coauthor of the merger, Brian Gilbertson. It should be noted here that in addition to Paul and Brian, Don Argus, who at the time headed the Board of Directors of BHP, also took an active part in the integration of the two companies (Thompson and Maclin, 2009). The merger of BHP and Billiton fully met the aspirations to create a new global mining company, recognized as an industry leader for its production and financial indicators. At the end of 2003, the combined company BHP Billiton was the leading global producer of coal and iron, a major producer of copper, aluminum, nickel, silver, zinc, and lead from mining and industrial companies on six continents. The company employed 35,000 employees, the total gross income exceeded US$10 billion dollars, and the net profit was at US$1.4 billion dollars. This was a completely different picture compared to the disastrous situation in 1998.

Brian Gilbertson replaced Paul Anderson as President of the company. Sales of the joint company went up to US$20 billion dollars and capitalization exceeded US$38 billion dollars.

Everything developed in the best possible way for the newly merged company. Financial indicators went up, and with them grew the ambitions of the new head of the company. Brian was a very aggressive player in the production of minerals and craved a stellar growth. With the new platform, he started thinking about major acquisitions. The company that was in his sights was none other than the global giant Rio Tinto.

When Don Argus learned about these plans, he had a strong adverse reaction to this initiative. Don, nicknamed "Don't Argue," still had not come to terms with negative feedback from the BHP shareholders following the acquisition of Magma. On January 6, 2003, during the meeting of the BHP Billiton Board of Directors, Brian Gilbertson was shown the door, after only a 6-month term as the President of the company. He subsequently announced his resignation, citing "insurmountable contradictions" with the Board of Directors. The Board of Directors then appointed, as a matter of urgency, Charles (Chip) Goodyear, who was at that time head of the company strategy development.

Chip worked closely with Paul Anderson and gained an excellent reputation in the company as a very practical "down-to-earth" leader. The news about this new appointment reached Chip while on a holiday in Idaho, from where he was summoned by the Board of Directors to Melbourne. As Chief Development Manager, Chip assisted CEO Paul Anderson in turning BHP into a stronger company making it less vulnerable to the financial decline the company suffered in the late 1990s. He saw that the success of mining company was not entirely dependent on cycles of commodity price fluctuations but also on the strength of the foundation of the company. He believed that top-to-bottom diversification in the range of company products, cultivating a secure network of clients and end-consumers, and by striving to increase its competitive edge are some of the critical underpinnings of a strong foundation. This should give the company a good chance for rapid growth, from which all shareholders were to receive substantial returns.

Chapter 13

Ivanhoe Achieving Full Ownership of Oyu Tolgoi

Chapter Outline

Imants Kavalieris—*1972 BSc graduate of University of Newcastle, NSW, Australia. He joined the Oyu Tolgoi team in September 2001. Imants was an experienced geologist with good knowledge on geology of copper*—*gold porphyry deposits based on his previous experience at Grasberg in Indonesia. Imants stayed at Oyu Tolgoi for almost 11 years until 2012 studying mineralogy and conditions of formation of Oyu Tolgoi deposits. Imants founded the Mineralogical Laboratory on the site where a number of young talented local geologists grew and developed their careers. Imants is currently a consulting geologist at Plus Minerals LLC in Mongolia.*

David Crane *earned a bachelor's degree at the University of Sydney in 1975, and a PhD at the University of New South Wales, Australia in 1978. Early in his career, Dave worked for the Goldfields Company in Indochina. In March 2001, he joined Ivanhoe Mines and worked on the Oyu Tolgoi program until September 2012. He held the position of senior geologist sharing the responsibility for the drilling program and geological mapping with the responsibility of establishing a database for*

Discovery of Oyu Tolgoi. https://doi.org/10.1016/B978-0-12-816089-3.00013-5

Oyu Tolgoi. Now retired from Ivanhoe, he is still active as a consultant from his home base Perth, Western Australia.

***Cyrill Orssich** is a 1981 BSc graduate of the Carleton University in Ottawa, Ontario, Canada. Cyrill is an exploration geologist with long career of international work and a proven record of success on projects from grass roots exploration to producing mines in Canada, South America, Philippines, Indonesia, Australia, Mongolia, Kazakhstan, Myanmar, and Burkina Faso. Cyrill has profound experience in porphyry Cu–Au–Mo, epithermal and orogenic gold, IOCG, Pb–Zn SedEx and W skarns. Cyrill worked at Oyu Tolgoi from February 2002 to 2013. Currently Cyrill Orssich is a consultant living in Squamish, British Columbia, Canada.*

***Dale A. Sketchley** graduated from University of British Columbia with BSc and MSc. Dale is an expert in resource geology QAQC in ore deposits at advanced exploration projects and mining operations. Before Ivanhoe, Dale worked at Placer Dome from 1998 to 1998 and then Acuity Management Ltd. He Worked at Oyu Tolgoi from June 2002 to April 2013. Currently Dale is at Acuity Geoscience Ltd in Vancouver, Canada. Dale conducts independent resource data evaluations for NI43-101 disclosures.*

*The energetic **Peter Terry**, a native Australian, was a specialist in "core teaching." He led a team of technicians responsible for systematically measuring and recording technical data from the drill core. These data were critically important to complement the geological core logs, described in an earlier section. Terry worked on the project until 2008.*

April 8, 2002 marked another turning point in the history of the Oyu Tolgoi project. On that day, BHP Billiton received letter from Ivanhoe Mines with the declaration that Ivanhoe Mines had fulfilled the terms of the Agreement with BHP, dated May 5, 2000, by completing an exploration program of the value of US$6 million at Oyu Tolgoi, which includes an expenditure of US$1.5 million dollars at Central Oyu Tolgoi.

According to the terms of the Agreement, BHP Billiton had 60 days from the date of receipt of the letter to verify and confirm that Ivanhoe Mines had satisfied all the commitments of the Agreement. These included the resource of chalcocite mineralization not less than 250 million tonnes with a copper content of 1% or higher or a resource of hypogene mineralization not less than 300 million tonnes with a copper content of 1% or higher. In the event that these conditions were met, BHP could come back into the project. Otherwise, it had to transfer the project to Ivanhoe. The deadline for BHP's response was June 7, 2002.

At the time of the audit, it was acknowledged that from May 5, 2000 to April 8, 2002 Ivanhoe Mines had completed drilling of 183 of diamond core holes and noncore reverse circulation (RC) drilling. Most of the drilling was performed at Central, South, and South-West Oyu (see Fig. 13.1). Besides drilling, various types of exploration work had been done, including deep-penetrating IP survey, hydrogeological studies, etc. The breakdown in terms of the type of drilling is presented in Table 13.1. To complete the drilling of the large number of holes that were planned during the earn-in process, Ivanhoe made the decision that as of August 1, 2001, drilling operations at Oyu Tolgoi would be conducted without interruption during the winter.

In addition to the amount of drilling, BHP also verified the costs incurred by Ivanhoe. According to the Agreement Ivanhoe provided a package of information to BHP Billiton for the performed exploration work. Based on this report, the consulting group of Roscoe Postle Associates prepared an NI 43-101 report in accordance with the requirements of the Canadian industry standard. To determine whether the resource tonnage requirements had been met based on the drilling data, the Amec Company produced a resource calculation for South-West Oyu and all of the Oyu Tolgoi prospects. Silvia Satchwell from BHP Billiton in Santiago and Zofia Ashby from Kilborn Engineering Pacific Ltd independently built their own models based on the data from Ivanhoe Mines.

BHP Billiton checked and validated the results of the resource calculation and determined that the resource modeling did not reach the previously established parameters for triggering the claw-back conditions. Therefore, there was no obvious reason for BHP Billiton to exercise the buy-back option. Following these conclusions, BHP Billiton initiated the transfer of the Oyu Tolgoi license to Ivanhoe Mines according to the terms and conditions of their Agreement. BHP Billiton then filed an application to MRAM, and the Government of Mongolia approved the request for the license transfer.

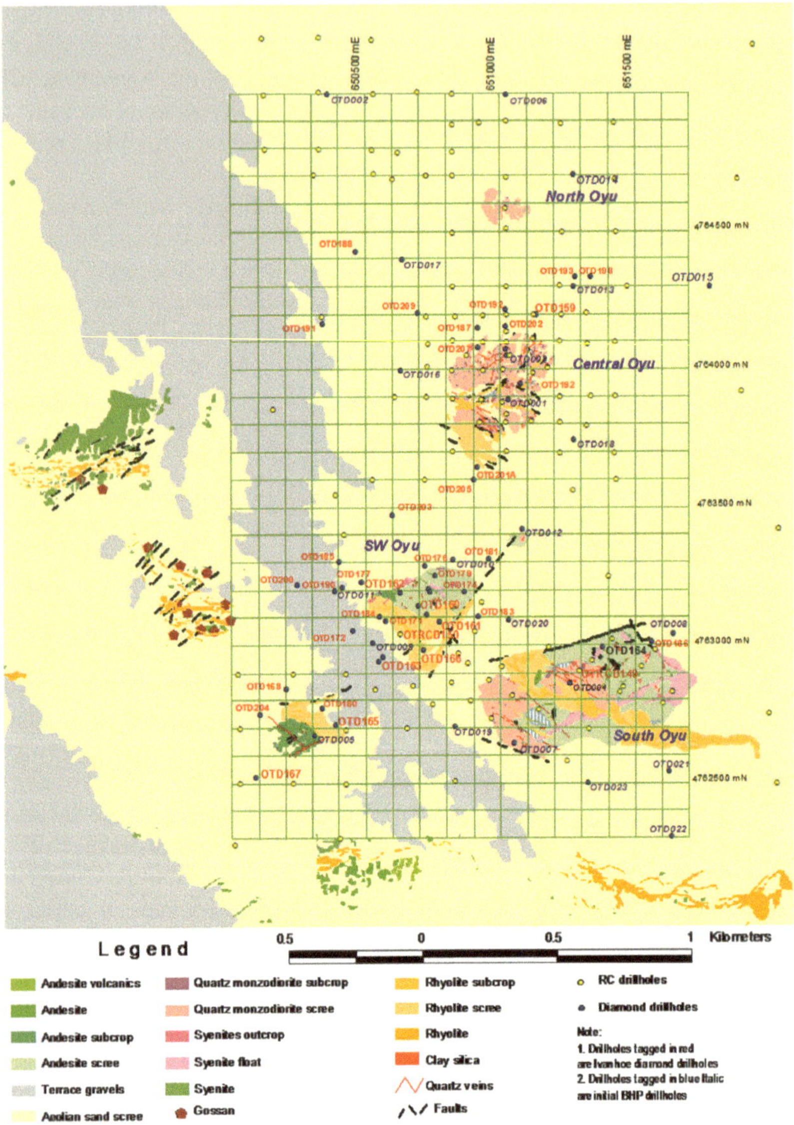

FIGURE 13.1 The geological map of Oyu Tolgoi showing Ivanhoe's diamond drill hole locations at Central, South, and South-West Oyu Tolgoi up to April 8, 2002.

Initially, Ivanhoe's priority was to identify additional resources of the near-surface secondary chalcocite blanket by RC drilling. After drilling dozens of shallow holes in May 2001, it became apparent that the secondary copper mineralization was restricted to Central Oyu Tolgoi. Charlie Forster calculated the resource to be in the order of 50 million tonnes, which was too small for

TABLE 13.1 Summary of Drilling Conducted by Ivanhoe Mines

Company	Type of Drilling	Number of Drill Holes	Amount of Drilling, Line-Meters
Ivanhoe Mines	Reverse circulation	133	11,870
Ivanhoe Mines	Combined	3	996
Ivanhoe Mines	Diamond core hole	47	26,797
Total		183	39,663

Ivanhoe in this remote environment. This was an unexpected setback and a critical stage for the Oyu Tolgoi project. Some leaders of Ivanhoe Mines began raising doubts regarding the project's prospectivity. The question in the minds of Forster's exploration team was "where to go next?"

For the remaining project budget money, Charlie decided to drill three deep diamond drill holes. He chose to drill Oyu OTD159 at Central Oyu and OTD149 at South Oyu. He hesitated a little with the selection of the third drill hole OTD150 and decided to turn to D. Garamjav for his advice. Garamjav's suggestion was to redrill BHP's borehole OT10 in which a 37-m interval of 1.2 g/t gold and 0.4% copper had been intersected. This hole had been terminated prematurely without reaching the targeted depth in a fault zone due to insurmountable drilling conditions. Garamjav then unfurled a schematic geological cross-section through drill holes OT07-OT10-OT12, which he had drawn up in 1998, and showed it to Charlie outlining the proposed location of the drill hole. Garamjav was both surprised and pleased that the head of the project supported his recommendation. By accepting Garamjav's proposal, Charlie committed to a deep drill test within the "golden triangle." He publicly stated that if this drill hole crosses the orebody, it would be Garamjav's discovery.

With the change of property ownership, destiny did not allow BHP the opportunity to eventually twin this hole. However, Ivanhoe now had the chance to follow up the modest intercept in OT10 and at the same time continue to check the validity of BHP's concept that the real prize at Oyu Tolgoi may be a Grasberg-style gold-enriched, hypogene copper porphyry deposit.

The collar of drill hole OTD150 was positioned on the opposite side of the fault zone from OT10 to avoid drilling into the same tectonic zone. This hole was drilled to a depth of 578 m and intercepted 508-m interval of mineralization with an average content of 1.17 g/t gold and 0.81% copper. The news about this intersection was sent from the field to Ivanhoe Mines corporate office in Vancouver via satellite. In the life of many discoveries, there is a critically important drill hole, which suddenly unlocks the mystery of the deposit and breathes new life into the project. Borehole OTD150 was this drill hole! The remaining drill holes at South-West Oyu Tolgoi were all deep.

In summer 2002, Ivanhoe management in Canada started looking for potential investors for the project. While three deep holes OTD149, OTD150, and OTD159 were being drilled, the representatives of some interested companies visited Oyu Tolgoi and saw some preliminary drilling results. The core from drill hole OTD159 with its display of long intervals of glittering bright blue covellite made a strong impression. The two other drill holes, containing intense quartz stockworks with abundant chalcopyrite, were equally impressive. D. Garamjav could not conceal his satisfaction over those drill intercepts, and one might say, he almost "completely lost his sleep" for excitement.

The answer to the question of "where to go next?" was to continue tracing the gold-enriched hypogene mineralization at South-West Oyu Tolgoi where BHP had values of copper and gold. There were also geologists in the Ivanhoe Group with experience in the copper–gold deposits, and they supported focusing the drilling on this mineralization style. Imants Kavalieris, who had previously worked at Grasberg in Indonesia, was one of them. Robert Friedland is known for his courage in making key decisions in risky situations and he supported the proposal from Charlie and Garamjav to drill several deep drill holes to further test the concept of hypogene mineralization.

At the end of 2002, about 6 months after BHP acknowledged the fulfillment of conditions for Ivanhoe's full ownership of the project, Ivanhoe initiated drill testing of the Far North Oyu Tolgoi prospect situated about 650 m north of North Oyu Tolgoi. This target consisted of a wide geophysical IP anomaly delineated by a recently conducted IP survey, which had a greater penetration depth compared to the survey that BHP had conducted previously. Ivanhoe covered the whole license area with a deep IP survey and altered the direction of the IP profile lines from South-North to East-West. The deeper penetration capability of the survey (allegedly down to the depth of 1,600 m) in combination with a more effective line orientation resulted improved the detection of deeply seated and broadly defined anomalies. The new IP map revealed that the anomalies were trending in an NNE-SSW direction.

Diamond drill hole OTD270 penetrated quaternary sediments exceeding a thickness of 200 m, underneath which a 638-m interval of primary chalcopyrite-bornite mineralization was intersected containing a high copper content. Having found significant tenor hypogene copper–gold mineralization at both the southern and northern ends of the Oyu Tolgoi license was a confirmation that this was indeed the priority target at Oyu Tolgoi. That was the trigger for escalation of the drilling program, and additional 13 drill rigs were transported to Oyu Tolgoi.

The most impressive intercepts at Far North Oyu Tolgoi came from drill hole OTD367A with 155 m averaging 4.41% copper and 1.61 g/t gold and from drill hole OTD409 with 66 m of 3.52% copper and 1.88 g/t gold.

A number of other drill holes yielded intercepts in excess 200 m with more than 3% copper and additional gold credits. The drill hole OTD150 was located in the so-called "golden triangle."

Some of the other Ivanhoe personnel who worked tirelessly in the background and contributed to the success of the exploration program at Oyu Tolgoi were David Crane, Cyrill Orssich, Dale Sketchley, and Peter Terry.

In addition to his geological skills, David Crane was also a good computer programmer. While working intensively in his field office at Oyu Tolgoi, David designed a software to process all of the Oyu Tolgoi geological data. The volume of data generated at Oyu Tolgoi turned out to be a rigorous test for his brainchild. At the beginning, his program often failed to cope up with the volume of data but David was imperturbable. With his creative mind, he was quickly able to make the necessary adjustments to rectify the problem.

David called this program GOD (Oyu Tolgoi Geological Database). It became the official Oyu Tolgoi field database. S. Sanjdorj jokingly christened David as Godfather of the Oyu Tolgoi database ("GOD Father"). Yes, he was considered to be some kind of genius to local geologists and indeed he made an invaluable contribution to the development of Oyu Tolgoi. David worked on the project until 2012.

After the analytical results had been interpreted, David assigned specific colors to the anomalous value ranges of the elements of interest. The high contents were in red, medium were in blue, and lower values were in green. When the first analytical results from drill hole OTD150 were arrived, the colors were predominantly red and blue. Charlie Forster and Ivanhoe's exploration manager Doulas Kirwin flew by helicopter from Ulaanbaatar to Oyu Tolgoi to rejoice and celebrate the occasion.

Cyrill Orssich, a 3D-modeling expert, worked on the models of the Oyu Tolgoi ore bodies using databases created by David Crane. His three-dimensional models made a significant contribution to the successful development of the geological exploration programs. Based on these models, it was always possible to visually determine where exploration should be directed to tracking of locate the extensions of the mineralization. Consequently, Ivanhoe also used his 3D models for calculating orebody volumes.

Qualitative and quantitative control of the sample preparations for chemical analysis, as well as the processing of their results, was managed by Dale Sketchley.

We should note here that many young local geologists worked hand-in-hand with Ivanhoe experts and as a result developed their extensive experience of conducting successful exploration. These include mineralogists Jargaljav Zandanshatar, Oyunchimeg Renchina, and Hashgerel Bat-Erdene and geologists Niislelhuu Gombosuren, Otgonbayar Togtohbayar, and Erdenebayar Togtoh. Ariunaa Tuvshintsengel worked under Dale's direction and took over his duties. Currently she is working at Oyu Tolgoi.

13.1 WATER RESOURCE FOR OYU TOLGOI

In the development of Oyu Tolgoi deposit, the discovery of groundwater deposits played a significant role. The lack of an adequate water supply in the project area was one of the critical risks for the success of the project. Among his other duties, geologist Samand Sanjdorj played a key role in the hydrogeology program.

Provisional estimates were that a future mining-concentrating plant at Oyu Tolgoi would require a water supply of 12,000 m^3/day or approximately 140 L/s. The mining-concentrating plant could, conceivably, be in operation for 50 years or more, and unfortunately, there were no big rivers that could meet the project's water requirements in the Gobi Desert. Therefore, the project would have to rely exclusively on underground water sources. Term of the plant activity could last for 50 years or more.

Water exploration primarily aimed at detecting large tectonic structures that could contain underground aquifers. The hydrologists knew that in this environment, only aquifers that were at least 300 m below the intermountain basins possessed sufficient porosity and saturation of groundwater to meet the demands of the future mine operation. Shallow aquifers were atmospheric precipitation and produced only relatively small volumes of water.

To detect such structures, the approach was to collect and study all data from previous hydrogeological surveys within a 100-km radius around Oyu Tolgoi. Gravity data and electrometric Vertical Electric Sounding data generated during petroleum exploration in these areas were also very useful in the search for water. Based on the data from the combined satellite images with geophysical information, the water explorers selected three promising sites—Gunii Hooloi, Galbyn Gobi, and Nariin Zagiin Hooloi. At each of these sites, TEM surveys were conducted. Based on the TEM sites more than a hundred water wells were sunk. Hydrogeologists conducted logging of each drill hole and performed trial pumping.

In 2001, at the beginning of the hydrogeological program, the Australian company Liquid Earth was subcontracted to undertake preparatory work, but then for some reason, they abandoned the project. In 2002, another Australian company Aquaterra was subcontracted to conduct hydrogeological prospecting. S. Sanjdorj's team conducted field geophysical TEM survey under the direct supervision of Dr. Brett Harris.

Following nearly 6 years of continuous hydrogeological exploration, two major deposits of underground water, Gunii Hooloi and Galbyn Gobi, were discovered. At the Gunii Hooloi deposit, with water supply of 870 L/s, a sizable water intake structure was built for future water supply to the processing plant at Oyu Tolgoi. The water intake structures consisted of about 30 deep wells (up to a maximum depth of 400 m), connected by a main tap line 90 km long. The engineers had to install two pumping stations for lifting water to a height of 100 m due to topographic relief from the source to the deposit and one closed-type tank with a capacity of 400,000 cubic meters of water.

13.2 DEDICATION TO MEMORY OF HUGO DUMMETT

In January 2003, during the Cordillera Roundup Conference in Vancouver, Sergei Diakov and Robert Friedland were chatting about the tragic and premature death of Hugo Dummett. Sergei proposed the idea of commemorating Hugo by naming the newly discovered Far North Oyu Tolgoi after him. Hugo was a great supporter of the Oyu Tolgoi project throughout the early stages and he strongly believed in its potential to become one of the world's major discoveries. Robert's reply was as follows: "I agree. We need to make sure that this is a major discovery." Later, in September 2003, Ivanhoe publicly announced that the deposit at Far North Oyu Tolgoi was named as the Hugo Dummett deposit. Douglas Kirwin in his comment indicated that the mine, which will be built at the Hugo Dummett deposit, would also bear his name Hugo Mine.

In his statement, Robert Friedland claimed that Ivanhoe's leadership was convinced that, in the end, the deposit would be in operation and it would take its rightful place among the ranks as one of the great deposits of copper and gold in the world. Mr. Friedland added that, when it became clear that the deposit was indeed a giant, it was worthy of Hugo's name who was truly a giant in the mining industry (Fig. 13.2).

FIGURE 13.2 Hugo Dummett with D. Garamjav on the top of the Javkhalant Mountain, photo September 2002.

The authors of this book, and those who had a chance to work directly under the leadership of Hugo Dummett, felt that naming the mine after Hugo was a fitting memorial in honor of him for future generations. He was the geologist, with whom we had openly discussed various issues that cropped up along the road toward discoveries not only in Mongolia but also in other parts of the globe. Hugo will always be remembered by us as our illustrious leader and as a visionary who devoted himself fully to the quest of making world-class discoveries. The Ekati diamond mine in arctic Canada was certainly a monument to Hugo as it was for Chuck Fipke and Stewart Blusson who discovered the new diamond province.

We, participants of the Oyu Tolgoi discovery, believe that his wife, Nora Dummett, his son, and daughter also share our opinion. We acknowledge the thoughtful gesture made by Ivanhoe Mines and its leadership, which heard our request and fulfilled their promises.

Chapter 14

BHP's Final Departure From the Project

In November 2003, Ivanhoe announced that it had reached an agreement with BHP Billiton to buy out the latter's 2% royalty on the project for US$37 million dollars. It is estimated that over a 20-year time span, the value of the 2% royalty would be almost US$900 million. Opting for a relatively modest amount of cash, instead of retaining the royalty, was a direct reflection on BHP's lack of confidence for the future of Oyu Tolgoi. However, an amount of US$37 million dollars would make an insignificant impact on the profit margin of the world's largest mining company such as BHP Billiton. Nevertheless, the company management felt that the cash plus the US$5 million from the sale of the project in 2002 was a decent income from Oyu Tolgoi considering that Discovery had only spent US$3 million making the discovery. In the past, the company corporate management frequently criticized the Discovery leadership for the destruction of shareholder value. However, the sale of the royalty at Oyu Tolgoi, in our view, was nothing other than a destruction of shareholder value. Retaining the royalty would have been the right decision in true value creation for the BHP Billiton shareholders.

Certainly, this was a very good financial arrangement for Ivanhoe Mines, and its CEO Robert Friedland cheerfully announced, "This is an excellent strategic acquisition for Ivanhoe Mines." The 2% royalty was the last remnant that BHP Billiton had of Oyu Tolgoi. In Ivanhoe's view and that of a leading group of independent experts, Ivanhoe had identified one of the world's largest high-grade copper and gold porphyry deposits. With 16 rigs drilling on the property in late 2003, it was noted that "the scale of the discoveries on Oyu Tolgoi continues to grow at a phenomenal rate as a result of intensive exploration program implemented through Ivanhoe drilling program." Friedland added, "Mongolia is becoming an attractive destination for global companies seeking copper and gold deposits. The number of major mining companies, following Ivanhoe's success with interest is increasing; some of

Discovery of Oyu Tolgoi. https://doi.org/10.1016/B978-0-12-816089-3.00014-7

them have already acquired licenses to prospective parts of Mongolia. With BHP Billiton out of the project, Ivanhoe had the freedom to choose an investment partner, whether it be an international company or a representative of the Mongolian government body, which will gain a significant share of future development."

The role of Oyu Tolgoi in the history of exploration in Mongolia is difficult to overestimate. We should pay a tribute to Ivanhoe Mines for their persistence and perseverance to succeed, for flexibility in adjusting the strategy models, for their courage to take risks, and for their ability to validate the concepts and achieve great results. It seems to us that this success was due to the company management, especially those line managers who strongly believed that their persistence in moving the exploration program in the right direction would ultimately lead to success. We also acknowledge the group of national geologists, including those that previously worked under the banners of Magma Copper and BHP, who worked on the project together with the specialists from abroad.

In case of Oyu Tolgoi, the roots of success go deeply into the framework of labor relations laid out by Magma Copper. We briefly describe them as PICT culture: perseverance, innovation, courage, and teamwork.

Successful exploration programs require a methodological and focused action plan, innovation, courage in decision-making, and perseverance. A superficial approach is unlikely to lead to success because it is easy to miss the subtle deposit indicators in the field.

It is obvious that the arsenal of outcropping deposits, particularly those with obvious Tier 1 size potential on the surface, is limited. Historically most of them have been detected and recorded by various methods. A new approach of capturing such data and its interpretation is paramount. Similarly, embarking on innovative ideas for data interpretation is equally critically important for the success of exploration discovery.

After interpreting the data, it is important to check the viability of accepted concepts of drilling. Deposits are discovered by drilling. Strict discipline is necessary in determining the priority of anomalous sites and verifying these anomalies by drilling. Courage to make decisions to drill test the right targets on the project is another important element of the successful discovery culture (Fig. 14.1).

Harmonious work of the geological teams is always a crucial factor contributing to successful discoveries. Hence, effectively working exploration

FIGURE 14.1 Oyu Tolgoi core review, February 2003. From left to right: S. Diakov, D. Garamjav, and T.O. Munkhbat

groups with strong teamwork environments is another essential prerequisite for successful discovery business.

How all these factors are materialized in the exploration team? It primarily depends on the leadership of the company and its representatives on the ground.

Chapter 15

Further Developments at Oyu Tolgoi

On August 18, 2004, Ivanhoe announced that AMEC (Australia-based AMEC Minproc currently subsidiary of AMEC Foster Wheeler), a reputable engineering company, prepared a new, independent review of the resource. The consultants estimated that Oyu Tolgoi contained a measured and indicated resource of 1.06 billion tonnes, grading 0.47% copper and 0.36 grams per tonne of gold (or 0.71% copper equivalent) at a 0.30% copper equivalent cutoff grade calculated for the mineralization down to the depth of 560 m. It also included the mineralization below 560 m at 0.6% copper equivalent cutoff. Altogether, the newly outlined mineralization contained approximately 11.2 billion pounds of copper and 12.4 million ounces of gold. At a higher 0.60% copper equivalent cutoff, the project included measured and indicated resources of 511.6 million tonnes grading 0.64% copper and 0.59 g/t gold or 1.00% copper equivalent containing approximately 7.2 billion pounds of copper and 9.3 million ounces of gold. Additional inferred resources of 1.2 billion tonnes, primarily contained in the Hugo Dummett deposits, grading 1.25% copper and 0.24 g/t gold (or 1.38% copper equivalent grade), at a 0.60% copper equivalent cutoff, hosting approximately 33.7 billion pounds of copper and 9.4 million ounces of gold (Turquoise Hill Press Release, August 18, 2004).

AMEC's estimate increased and upgraded the near-surface resources contained in the project's Southern Oyu Tolgoi deposits, which included Southwest Oyu Tolgoi, Far Southwest Oyu Tolgoi, Central Oyu Tolgoi, South Oyu Tolgoi, and the recently discovered Bridge, Wedge, and Southern Sliver zones. The deposits lie at the southern end of a series of cogenetic copper and gold deposits delineated to date in the 5.3-km trend of the mineralization of Oyu Tolgoi (Turquoise Hill Press Release, August 18, 2004).

In November 2004, Ivanhoe announced that ongoing step-out drilling on the Hugo North deposit at Ivanhoe's Oyu Tolgoi copper–gold project in Mongolia delineated what is believed to be the highest grade copper mineralization ever found in a porphyry setting anywhere in the world. The recent drilling also extended the length of the Hugo North high-grade copper–gold core to more than 1.6 km, a further 300 m beyond the discovery's northern limit that had been previously established (Turquoise Hill Press Release, November 18, 2004).

Discovery of Oyu Tolgoi. https://doi.org/10.1016/B978-0-12-816089-3.00015-9

The Hugo North deposit, as defined by the AMEC 2004 resource model, increased in tonnes and grade from south to north. At the northern end of this resource model, there was a 50-m thick, subvertical block. It contained approximately 15.5 million tonnes of mineralization at an average grade of 3.0% copper and 0.6 grams of gold per tonne calculated at higher than 2% cutoff. With lower cutoff at 1%, it contained nearly 27 million tonnes of 2.2% copper and 0.5 g/t gold. Based on the most recent results, the potential for higher grade mineralization significantly increased. Further drilling was required to track the mineralization along strike to the north before it can be defined as a mineral resource.

"The remarkable growth of the high-grade core of Hugo North holds very significant implications for our modeling and the economics of future underground mining design," Robert Friedland commented. "While these drilling results speak for themselves, the objective of the current in-fill drilling program is to define that 'sweet spot' in the deposit where initial, large-scale block-cave production can access the highest grades. The step-out drilling is adding a very large tonnage of the highest grade copper–gold mineralization encountered to date" (Turquoise Hill Press Release, November 18, 2004).

Previously, on November 10, 2004, both Ivanhoe Mines and Entrée Gold announced that they reached an earn-in agreement to allow Ivanhoe Mines to conduct deep exploration at the northern extension of Hugo Dummett into Entrée's ground at Shivee Tolgoi (Turquoise Hill Press Release, November 10, 2004). Under the terms of the joint venture between Ivanhoe and Entrée Gold, Ivanhoe Mines was preparing to use its deep-penetrating Induced Polarization (IP) systems and deep-hole diamond drilling to test the potential of the northerly extension of the Hugo deposit beyond the boundary of Ivanhoe's Oyu Tolgoi block.

On December 6, 2004, Ivanhoe announced that the company received encouraging results from the first pass of its proprietary, deep-penetrating IP survey on the Ivanhoe–Entrée Gold Inc. joint-venture property, contiguous with and directly north of Ivanhoe's 100%-owned Oyu Tolgoi Project. The preliminary IP survey results indicated that the Hugo North IP anomaly extended onto the Ivanhoe/Entree joint-venture property for approximately 4 km in a north-northeasterly direction. The IP survey also identified several other significant geophysical targets on the joint-venture property. The proprietary IP system used at Oyu Tolgoi property was developed by Delta Geoscience of Canada to explore for and delineate sulfide mineralization to depths not usually detected by conventional IP surveys. At Oyu Tolgoi, the IP surveys outlined a continuous chargeability anomaly over a 6-km strike length representing sulfide mineralization that hosts the Southwest Oyu, South Oyu, Central Oyu, and the Hugo Dummett copper and gold deposits. The northernmost 1.6-km section of the IP anomaly on the Oyu Tolgoi property represented the gold-rich, high-grade copper Hugo North porphyry deposit (Turquoise Hill Press Release, December 6, 2004).

Robert Friedland's speech on March 5, 2005, at the Investor Conference in Tampa, Florida sparked a violent reaction in Mongolia (Resource Investor, 2005). The original intention was quite positive. To advertise the attractive investment climate in Mongolia, Robert praised the approaches of the country's leadership toward mining legislation with the aim of creating favorable opportunities for mining companies. However, his choice of words for the comparison of favorable investment atmosphere in the country was unfortunate. To illustrate his point he compared mining investing with a five dollar investment in a T-shirt factory that would yield a profit of $100. The public interpreted this as "exploitation" of mineral and human resources. Friedland's speech had a few of hyperboles and exaggerations; for example, he wrongfully quoted the income tax rate in Mongolia to be as low as 5%. The statements about the absence of nongovernmental organizations and local population in the Gobi were also exaggerated. Instantly, this misinterpretation went viral through the Internet. It caused protests from nongovernment organizations and citizens of Mongolia demanding the legislators to review the laws of the country. As a reference, they used the situation with Erdenet. Indeed, the Soviet Union spent huge capital for the mining plant construction, and the Mongolian side in the joint venture received a controlling share of 51%. Of course, such examples from the Soviet past were not applicable to the contemporary realities of the market economy in Mongolia. However, some opponents of the most progressive mineral resource laws in the country used it as a pretext to attack the Government and its policy. Intensive criticism of the Law on Subsoil Use and Investment Law in Mongolia began. Even a new public movement called "50 to 50" started requesting 50% ownership split between the deposit discoverer and the State.

It was not surprising that Ivanhoe Mines immediately became the focus of everyone's attention in Mongolia. The shadow of mistrust was also falling onto the numerous other small Canadian junior companies, most of which, by the way, worked very hard toward the development of mineral resources of Mongolia. Moreover, these companies undertook an enormous burden of risk due to the uncertainty of the investment regime in the country at the time. Demands that only big mining companies with established reputations could manage large mineral projects started to escalate. Demands were even made that similar to Erdenet the controlling share in the large mining projects be given to the State. Hotheaded critics ignored the simple fact that small companies also play an essential role in the exploration process in the development of the Mongolia's mining sector. Thanks to the progressive law on foreign investment and mineral deposits, Mongolia, previously virtually unknown to the global mining sector, quickly became one of the top five leading countries in the world for exploration expenditures. However, the violent reaction to the speech in Tampa started to cast an unfavorable light on the mining companies across the nation.

Ivanhoe Mines now felt compelled to urgently find a major mining company interested in becoming a partner on the Oyu Tolgoi mine development project. Ivanhoe immediately turned their eye on BHP Billiton because the two companies were equal partners in an exploration program immediately north of Oyu Tolgoi. BHP Billiton brought its new proprietary Falcon airborne gravity geophysical system to the program. This was a rapid exploration tool to survey this large land package. BHP Billiton quickly agreed to consider the option of reacquiring a stake in Oyu Tolgoi. The project was much more palatable to BHP Billiton Discovery group now because with the discovery of Hugo Dummett ore body, it was heading toward a Tier 1 size copper–gold deposit. However, approval was still required from senior company management.

In August 2005, a technical team consisting of Diego Hernández, the President of BHP Billiton's copper business, Marcus Randolph from corporate business development, and Mike Salamon from senior management came to Mongolia to review the project. Ivanhoe's senior management was represented by Robert Friedland. The meeting of the technical team took place at Oyu Tolgoi (Fig. 15.1). All technical and economic data for the newly discovered resources at Oyu Tolgoi, particularly the Hugo Dummett deposit, indicated that the recently discovered ores with high gold, copper content elevated Oyu Tolgoi into a higher rank. The consensus of the BHP Billiton senior management team was that Oy Tolgoi now appeared to be a first-class (Tier 1) copper–gold porphyry deposit, and the terms proposed by Ivanhoe Mines were attractive.

On this basis, the BHP Billiton technical team prepared a proposal to the management of the company with the suggestion to acquire a stake in Oyu

FIGURE 15.1 Project review by BHP Billiton delegation, photo August 2005. From left to right: Diego Hernandez (BHP Billiton), Robert Friedland (Ivanhoe Mines), Peter Leaman (BHP Billiton), Charlie Forster (Ivanhoe Mines), Mike Salamon and Marcus Randolph (BHP Billiton).

Tolgoi and the proposal was presented to the Board of Directors of the company and CEO, Charles Goodyear, who had already articulated his intentions to leave his position. During the discussions of the proposal at the Board level, some doubts were expressed about the prudence of dealing with Ivanhoe Mines based on the concerns from Ivanhoe's marketing of Voisey's Bay nickel deposit in Canada in the 1990's (J. McNish, 1998). Ultimately, the BHP Billiton Board rejected the proposal. All of the many participants who had persevered and worked so diligently in an attempt to "land the big fish" for BHP and then for BHP Billiton were greatly disappointed. They were especially disheartened because now that it was established that Oyu Tolgoi had the credentials of a Tier 1 deposit, which by definition satisfied the company's requirements, the project was once again rejected but this time, seemingly, for personal reasons. It was clear that BHP Billiton had closed the door on Oyu Tolgoi once and for all, thereby rejecting the rare opportunity to acquire a world class deposit.

Rio Tinto, one of the three largest mining companies in the world, entered Mongolia in early 2000 looking for coking coal investment opportunities. Their search gravitated to the southwest Gobi. Soon, however, Rio Tinto was attracted to the Paleozoic copper porphyry belt in South Gobi. In October 2004, Rio Tinto entered into an agreement with Entrée Gold, a Canadian junior company, which owned exploration licenses at Oyu Tolgoi North, called Shivee Ovoo in the north and Javhalant in the south. In June 2005, Rio Tinto implemented the terms of the agreement, gaining 11% of the Entrée Gold company shares and, thus, opening the door for their entry into Oyu Tolgoi. It was an obvious move, pointing to the actual reason for coming to the Mongolian Gobi.

Soon after that, in October 2006, Rio Tinto made an agreement with Ivanhoe Mines to finance the project, thus creating a defined path for Rio Tinto to become the largest shareholder in Ivanhoe Mines (Turquoise Hill Press Release, October 18, 2006). A joint technical committee was established to manage the Oyu Tolgoi Project. The two companies agreed to cooperate in the construction and operation of Oyu Tolgoi. For US$1.5 billion Rio Tinto acquired a 20.8% stake in Ivanhoe Mines. In doing so, Rio Tinto was allowed to continue acquiring shares of Ivanhoe over the next 5 years, which increased its ownership to 40%. Rio Tinto CEO Tom Albanese stated in a press release acknowledging the establishment of its interest in one of the largest undeveloped copper–gold resources in the world (Turquoise Hill Press Release, October 18, 2006). The next step was to conclude an agreement with the Mongolian Government on a long-term development project.

Chapter 16

Oyu Tolgoi Resource Expansion

As intense drilling continued at Oyu Tolgoi, its resource base was expanding. The latest exploration had increased the inferred resource of the northern-most portions of the Hugo North Deposit, which was continuing to grow by ongoing drilling at Oyu Tolgoi. On February 1, 2006, Ivanhoe announced that a new mineral resource estimate prepared by AMEC amounted to more than 1.25 billion tonnes of 1.04% copper and 0.24 g/t gold, or 1.2% copper equivalent (Tables 16.1 and 16.2).

At the same time, exploration drilling continued at the Heruga Deposit. On March 12, 2008, Ivanhoe issued a press release with the first resource estimate for the Heruga copper, gold, and molybdenum deposit (Turquoise Hill Press Release, March 12, 2008). The Heruga Deposit adjoins the south boundary of Ivanhoe Mines' Oyu Tolgoi project. The deposit is located within the Javhalant license, which is part of the Entrée Gold-Ivanhoe Mines joint-venture agreement area. The Heruga deposit is estimated to contain an inferred resource of 760 million tonnes grading 0.48% copper, 0.55 g/t gold, and 142 ppm molybdenum, for a copper equivalent grade of 0.91%, using a 0.60% copper equivalent cutoff grade. Based on this initial estimate, the Heruga deposit was estimated to contain at least 8 billion pounds of copper and 13.4 million ounces of gold. At a higher cutoff grade of 1% copper equivalent, the Heruga deposit included inferred resources of 210 million tonnes grading 0.57% copper, 0.97 g/t gold, and 145 ppm molybdenum, totaling 2.6 billion pounds of copper and 6.4 million ounces of gold (see Tables 16.3 and 16.4)

With the new Heruga deposit discovery, the Oyu Tolgoi total resource estimate including measured, indicated, and inferred categories was more than 3.5 billion tonnes (Table 16.5).

After lengthy discussions and negotiations on the Feasibility Study, Mongolian Parliament decided to empower the Government of Mongolia to conclude the investment agreement. On October 6, 2009, Ivanhoe Mines and Rio Tinto signed a long-term investment agreement with the Government of Mongolia establishing a comprehensive framework for the construction and operation of the Oyu Tolgoi copper–gold mining complex. The agreement created a foundation for the partnership between the Mongolian Government, which under the agreement were to acquire a 34% interest in Oyu Tolgoi's

Discovery of Oyu Tolgoi. https://doi.org/10.1016/B978-0-12-816089-3.00016-0

TABLE 16.1 Grand Total Mineral Resources of Oyu Tolgoi as of January 2006[a]

					Contained Metal		
Resource Category	Tonnage (tonnes)	Cu (%)	Au (g/t)	Cu eq. (%)	Cu ('000 lbs.)	Au (ounces)	Cu eq. ('000 lbs.)
Measured	101,590,000	0.64	1.10	1.34	1,430,000	3,590,000	3,000,000
Indicated	1,046,970,000	1.34	0.42	1.60	30,800,000	14,060,000	36,950,000
Measured + Indicated	1,148,560,000	1.27	0.48	1.58	32,230,000	17,650,000	39,950,000
and + Inferred	1,250,550,000	1.04	0.24	1.20	28,750,000	9,880,000	33,070,000

[a] *Oyu Tolgoi mineral resources are inclusive of mineral reserves. Turquoise Hill press release, February 1, 2006.*

TABLE 16.2 Mineral Resources of the Ivanhoe/Entrée Shivee Tolgoi Joint Venture Property[a]

					Contained Metal		
Resource Category	**Tonnage (tonnes)**	**Cu (%)**	**Au (g/t)**	**Cu eq.(%)**	**Cu ('000 lbs.)**	**Au (ounces)**	**Cu eq. ('000 lbs.)**
Inferred	190,160,000	1.57	0.53	1.91	6,590,000	3,240,000	8,010,000

[a]*Turquoise Hill press release, February 1, 2006. Resource as of January 2006 at 0.6% copper equivalent cutoff grade.*

TABLE 16.3 Mineral Resources at Hugo North and South Deposits on the Oyu Tolgoi Property as of January 2006[a]

					Contained Metal		
Resource Category	Tonnage (tonnes)	Cu (%)	Au (g/t)	Cu eq. (%)	Cu ('000 lbs.)	Au (ounces)	Cu eq. ('000 lbs.)
Hugo North Deposit							
Indicated	581,330,000	1.91	0.41	2.17	24,440,000	7,620,000	27,770,000
Inferred	671,720,000	1.11	0.34	1.33	16,450,000	7,320,000	19,650,000
Hugo South Deposit							
Inferred	490,330,000	1.05	0.09	1.11	11,380,000	1,390,000	11,990,000
Total Hugo Deposits							
Indicated	581,330,000	1.91	0.41	2.17	24,440,000	7,620,000	27,770,000
Inferred	1,162,050,000	1.08	0.23	1.24	27,830,000	8,710,000	31,640,000

[a]*Turquoise Hill press release, February 1, 2006. Copper equivalent calculated at 0.6% cutoff.*

TABLE 16.4 Inferred Resources of the Heruga Deposit as of March 2008[a]

						Contained Metal		
Cu eq. (%) Cutoff	Tonnage (tonnes)	Cu (%)	Au (g/t)	Mo (ppm)	Cu eq. (%)	Cu ('000 lbs.)	Au (ounces)	Cu eq. ('000 lbs.)
>1.00	210,000,000	0.57	0.97	145	1.26	2,570,000	6,400	5,840,000
>0.60	760,000,000	0.48	0.55	142	0.91	8,030,000	13,400	15,190,000

[a]*Turquoise Hill press release, March 12, 2008.*

TABLE 16.5 Total Oyu Tolgoi Project Resources[a]

						Contained Metal		
Resource Category	Tonnage (tonnes)	Cu (%)	Au (g/t)	Mo (ppm)	Cu eq. (%)	Cu ('000 lbs.)	Au (ounces)	Cu eq. ('000 lbs.)
Measured	101,590,000	0.64	1.10	–	1.34	1,430,000	3,590,000	3,000,000
Indicated	1,285,840,000	1.38	0.42	–	1.65	39,120,000	17,360,000	46,770,000
Measured + Indicated	1,387,430,000	1.33	0.47	–	1.63	40,680,000	20,970,000	49,860,000
Inferred	2,157,130,000	0.80	0.35	50	1.05	38,230,000	24,220,000	50,050,000

[a]*Turquoise Hill press release, March 12, 2008.*

license holder, Ivanhoe Mines Mongolia Inc. According to this Agreement, Ivanhoe Mines was to retain a controlling 66% interest in Oyu Tolgoi. Global miner Rio Tinto, which joined Ivanhoe Mines as a strategic partner, was holding a 9.9% interest in Ivanhoe Mines. After several years, Rio Tinto intended to increase its stake to 51% of Ivanhoe shares.

The Mongolian Parliament authorized the Government to finalize the investment agreement through a special resolution approved on July 16, 2009. The agreement guaranteed a 50-year assurance of stability for what was expected to be a 100-year mine life based on the mineral resource of Oyu Tolgoi. A 2009 revised resource estimate was that Oyu Tolgoi contained approximately 79 billion pounds of copper and 45 million ounces of gold in measured, indicated, and inferred resources (Turquoise Hill Press Release, October 9, 2009).

From that point onward, Rio Tinto's share position in Oyu Tolgoi increased rapidly. At the end of 2009, the Ivanhoe Mines, Rio Tinto Oyu Tolgoi Technical Committee approved a conditional US$ 758 million budget for 2010 to begin full-scale construction of the copper–gold mining complex in southern Mongolia. The construction of the Concentrator and open-pit mine was completed in 2012, and the commercial production began in 2013.

Ivanhoe began a new exploration program to drill test previously unexplored areas around the deposits along strike and at depth. Measured and indicated resources at Oyu Tolgoi now totaled approximately 1.4 billion tonnes at an average grade of 1.33% copper and 0.47 g/t gold, plus an additional 2.4 billion tonnes of inferred resources at an average grade of 0.78% copper and 0.33 g/t gold. The estimated mineral reserves are a subset of these resources totaling 1.39 billion tonnes (Turquoise Hill Press Release, May 11, 2010).

On September 28, 2010, Ivanhoe announced that the company had intersected almost 1 km of near-continuous gold and copper mineralization in drill hole OTD1510 at Oyu Tolgoi, making it the longest exploration drill intercept of gold and copper mineralization recorded since Ivanhoe began drilling at the Oyu Tolgoi Project in 2001. This was after drilling 1,650 drill holes with the total length of drilling 900 km (Turquoise Hill Press Release, September 8, 2010).

The area, previously known as the New Discovery Zone, has now been renamed Heruga North. The OTD1510 intercept indicated that Heruga North is part of a 2.5-km, gold-rich mineralized extension of the Heruga deposit, stretching north from the southern border of the Oyu Tolgoi mining license to the southern Oyu Tolgoi deposits. Based on interpreted geology and a significant, coincident, gradient-array IP anomaly identified by the proprietary, deep-exploration technology, drill hole OTD1510 targeted a critical 600-m gap in the known mineralization between the northern, fault-controlled limit of the Heruga deposit and the Heruga North deposit. The Heruga North discovery confirmed potential for further expansion of the Oyu Tolgoi system. The 938-m Heruga North intercept in drill hole OTD1510 covered a horizontal distance of 643 m and a vertical distance of 681 m. Altogether, Ivanhoe had completed approximately 43,500 m of wide-spaced diamond drilling at the Heruga North zone.

A March 2010 estimate indicated that the Heruga deposit contained 10.2 billion pounds of copper and 15 million ounces of gold. The gold-rich zone at Heruga North was analogous to the gold-rich zone at the Heruga deposit. "Based on the high gold and copper grades, as well as the high gold-to-copper ratio and the style and tenor of the porphyry mineralization, the significance of Heruga North compares favorably to the major deposits that currently comprise the Oyu Tolgoi Project," indicated by R. Friedland (Turquoise Hill Press Release, September 8, 2010).

In January 2012, Rio Tinto ownership of Oyu Tolgoi reached 51% of opening a new era for the project under the new banner of Rio Tinto.

Chapter 17

Oyu Tolgoi—New Major Copper Producer in Mongolia

The year 2013 was historic for Oyu Tolgoi, the first ever copper–gold production commenced. On July 8, 2013, Ivanhoe announced that the Oyu Tolgoi mine began shipping copper concentrate. The initial sale of approximately 5,800 tonnes of concentrate was sent to customers in China (Turquoise Hill Press Release, July 8, 2013).

After achieving the milestone of shipping 2.5 million tonnes of concentrate in October 2016, the Turquoise Hill team was rapidly advancing toward the next milestone of 3 million tonnes in June 2017. Oyu Tolgoi continues to grow its production capacity, and it is becoming a critical part of the Mongolian economy.

On October 16, 2017, Turquoise Hill announced that during the third quarter, Oyu Tolgoi set three operational records for total material mined, ore treated, and average daily concentrator throughput (Turquoise Hill Press Release, October 16, 2017). Copper production for the quarter was in-line with the second quarter, whereas gold production increased by almost 30%. Production in the third quarter for the material mined increased by 9.0% compared to the second quarter and amount of treated ore increased by 10.1%. The average daily concentrator throughput during the third quarter was 115,400 tonnes, which was 8.9% higher than in the previous quarter. Copper production was mostly flat, whereas gold production increased by 29.2% due to higher head grades from the medium-grade stockpile. Higher gold grades during the third quarter of 2017 resulted in a 26.1% increase in gold sales compared to the second quarter.

The dream of establishing a new copper production center in the south of Mongolia came true. Oyu Tolgoi has become a significant copper producer in Asia and the world and will continue to be so in the foreseeable future. Modern technology has arrived in the Gobi Desert with the development of a network of paved roads and with new power lines bringing electrical energy to the region (Fig. 17.1). We hope that new power plants will be built very soon, converting this part of Mongolia into a self-sustainable world-class copper producing district (Table 17.1).

Discovery of Oyu Tolgoi. https://doi.org/10.1016/B978-0-12-816089-3.00017-2

FIGURE 17.1 Overview of the Oyu Tolgoi production facility operated by Rio Tinto. *Photo by Sanjdorj April, 2018.*

TABLE 17.1 Oyu Tolgoi Quarterly Production Data in 2016 and 2017[a]

	3Q 2016	4Q 2016	1Q 2017	2Q 2017	3Q 2017	9 Months 2016	9 Months 2017	Full Year 2016
Open pit material mined ('000 tonnes)	25,739	25,615	24,333	25,193	27,466	71,322	76,992	96,938
Ore treated ('000 tonnes)	9,146	9,819	10,087	9,637	10,615	28,333	30,339	38,152
Av. Mill Head Grades								
Copper (%)	0.66	0.61	0.51	0.51	0.48	0.67	0.5	0.65
Gold (g/t)	0.21	0.25	0.15	0.16	0.18	0.39	0.16	0.36
Silver (g/t)	1.99	1.5	1.3	1.38	1.34	1.95	1.34	1.83
Concentrates produced ('000 tonnes)	203.2	206.7	176	171	170	639.8	517	846.6
Average concentrate grade (% Cu)	22.9	22	21.6	21.8	21.7	24.4	21.7	23.8

TABLE 17.1 Oyu Tolgoi Quarterly Production Data in 2016 and 2017[a]—cont'd

	3Q 2016	4Q 2016	1Q 2017	2Q 2017	3Q 2017	9 Months 2016	9 Months 2017	Full Year 2016
Production of Metals in Concentrates								
Copper ('000 tonnes)	46.6	45.5	38.1	37.2	36.9	155.9	112.1	201.3
Gold ('000 ounces)	37	49	25	24	31	251	80	300
Silver ('000 ounces)	361	273	215	236	239	1,147	689	1,420
Concentrate sold ('000 tonnes)	206.2	182	190.2	182	176.6	646.6	548.8	828.6
Sales of Metals in Concentrates								
Copper ('000 tonnes)	45.7	37.6	39.5	37.3	36.9	151.3	113.6	188.9
Gold ('000 ounces)	38	39	32	23	29	307	84	347
Silver ('000 ounces)	341	239	205	222	229	1,041	656	1,280
Metal Recovery (%)								
Copper	78	76.6	74.9	74.6	73.5	82.4	74.3	81
Gold	62	63.4	48.8	47.7	51.2	69.7	49.4	68.5
Silver	61.7	57.2	51.8	53.9	52.8	64.7	51.8	63.1

[a]*All data represents full production and sales on a 100% basis.*

Oyu Tolgoi has created more than 13,000 jobs, 94% of which are held by Mongolians (Turquoise Hill BMO Global Mining presentation. February 2017). Many thousands of additional jobs directly or indirectly servicing Oyu Tolgoi copper production were created in the country. Oyu Tolgoi has given a major boost to the growth of the Mongolian economy.

Chapter 18

Conclusion

At the end of our story, we note that there are useful lessons to be learned from the historical events of the Oyu Tolgoi discovery. We attempted to highlight some of them in this book, but we also leave some room for the observant readers to make their own conclusions and interpretations.

Since Ivanhoe's involvement with Oyu Tolgoi project, considerable information had been disseminated to the public mainly through press releases. However, apart from documentation by Perelló et al. (2001), Ochirbat (2010), and Sanjdorj (2011) and in the more recent book by Garamjav (2014), which touched on a number of aspects of the early phase of the Project, the full pre-Ivanhoe story was untold until the publication of this book.

Indeed, early stages of Oyu Tolgoi were critical to the future life of this significant discovery in the global mining industry. The details of how the concept of exploring for copper porphyry deposits in Mongolia was born, what led a band of mineral explorers assembled from various parts of the globe to the southern depths of the Gobi Desert in search of the target, and what accompanied the first triumph of the discovery of this giant Oyu Tolgoi deposit would help those interested in acquiring a complete picture of this story. It is our firm belief that the events at the early stages had a profound effect on the successful outcome at Oyu Tolgoi.

The general framework was vitally important. Firstly, Mongolia managed to create a favorable investment climate in the country by adjusting their legislation on foreign investments and subsoil minerals. The government accomplished this by introducing a very efficient mineral licensing system. These legislative acts had a potent stimulatory effect on the geological exploration environment, which in turn contributed to a sharp increase in the number of investments for mineral exploration, eventually leading to discoveries of new deposits in Mongolia.

The approach adopted by the Erdenet-Magma Joint Venture team and then by BHP followed the general principles of the systematic approach: from general to the particular. Initially, our task was to find a new significant copper deposit in Mongolia. A team of handpicked specialists developed an exploration strategy focused on the JV goal of a major copper discovery for both

Discovery of Oyu Tolgoi. https://doi.org/10.1016/B978-0-12-816089-3.00018-4

partners. Initially, we took a broad approach of selecting the various genetic types of copper deposits that could have the potential of yielding the desired copper discovery in the geologic conditions of Mongolia. However, in the course of the validation of copper occurrences of different genetic types, it became clear that only a porphyry type of copper mineralization enhanced by either a zone of secondary sulfide enrichment or with a significant by-product such as gold could yield the right outcome. For each type of copper deposit, we selected appropriate sets of prospecting and exploration methods, which occasionally had to be adjusted.

We applied modifications and adjustments not only to the set of selected search methods but also to the ultimate goals of our exploration, which also evolved during the course of the program. If the first joint venture Erdenet-Magma sought a copper porphyry deposit on the scale and content of the metal comparable with Eredentuin Ovoo, then, during the BHP era, the size threshold was increased to Tier 1 or world-class copper deposits. The increase in target size was particularly challenging because highly selective approaches were required to identify only those geological settings that could contain giant deposits.

Effective teamwork is a critical element in a successful mineral exploration program. Here, the Magma Copper culture played an essential role. Teamwork was based on the trust that each member fulfilled their assigned duties responsibly, that they could count on each other's support unconditionally, and the faith they had in the ability of the team to achieve successful results. This type of team spirit empowered the team with vigor and a thirst for success. The open atmosphere for discussion of ideas, continuous search for the most efficient solutions allowed the team to define realistic goals, effortlessly adjusting them as needed. We collectively called these elements as PICT or successful discovery culture that comprises such ingredient features as persistence, innovation, courage, and teamwork.

Selection of the team staff was a significant enabling factor. The international character of geological exploration teams did not prevent extensive exchanges of views between the experts. Good collaboration gave an opportunity for the program to conduct and complete a significant amount of reconnaissance work in a short period, despite extremely challenging conditions of the fieldwork. Both Magma Copper and then BHP relied on national staff who played a pivotal role in achieving the success.

Professional openness and trust in the discussions allowed the team to conduct profound analysis and make rational choices for the prospective areas with the best potential for future discoveries. This, in many ways, enabled the performers of the program to relatively quickly identify the path to success, concentrate on the right targets, and conduct their work in the best efficient manner.

At each stage, the program management identified the goals, determined the scope of work, and laid out approaches toward successful achievement. An intensive study of the geological material from published sources and available state-owned geological library reports in the country preceded the beginning of the fieldwork. With the concepts generated in the process of the original data review, the project geologists were able to quickly identify and select the areas of interest for their field validation work focusing their attention on the most prospective ground.

Discoveries usually begin in the offices as desktop studies or discussions of various ideas, which have to be verified in the field. The ultimate test of these ideas is drill testing. The types of mineral deposits determine selection of types and scales of prospecting and exploration activities. Large deposits usually require adequate size of the exploration grid, which explorers should apply in their search efforts. Appropriate choice of the exploration grid size allows explorers, on the one hand, to detect the right size of target and, on the other hand, help them to avoid unnecessary costs early on. The size of the exploration "mesh" must match the size of the target allowing explorers to catch the right size of "fish." The issues of adequate selection of horizontal exploration grid at an early stage should link the problems of vertical zoning and distribution of mineralization at depth. Geological models predetermine the size of exploration grid, but early on, one needs to exercise flexibility in the use of the selected models. Focus on one model without a full understanding of the deposit may result in an inadequate evaluation of the deposit potential.

New approaches to the data interpretation, adjustment of models, and searching methods gave the team an opportunity to direct their program in an optimal way to achieve the ultimate goal. The expertise from other regions enabled the team quest to review and apply a wide range of possibilities based on well-known patterns. Of course, the flipside of the coin was that due to the rigid attachment to specific models the project had to pay its price—the real potential of the project hypogene mineralization was not recognized until later on. Nevertheless, the most critical aspect of the whole process at an early stage of Oyu Tolgoi discovery remained consistent. It was based on discipline for objective data collection in combination with flexibility in its interpretation.

We used a full arsenal of different exploration tools in our search methods: remote sensing with Landsat and SPOT images, analysis of volcano–plutonic formations in combination with regional structural analysis to identify the centers of volcanism and related hydrothermal activity capable to generate mineralized porphyry systems of ore district scales. Within these ore fields, we applied a set of additional search tools including geophysical, geological, geochemical, and mapping methods allowing us to delineate anomalous areas for future drill testing (Fig. 18.1).

Deposits become first-class in the evolution of exploration. Along this journey, they transition from ordinary mineral manifestations or occurrences, gradually advancing from small to large deposits through the exploration stages. Only then, they become Tier 1 discoveries, provided they possess such geological potential. Another catch here is that spending too much time and

FIGURE 18.1 BHP Billiton delegation visit to Oyu Tolgoi in September 2001. Above left to right: Craig Panther, Tumur-Ochiriin Munkhbat, and Sergei Diakov. In the middle: Dondogiin Garamjav. Below left to right: Jack McClintock, John-Mark Staude, and Douglas Kirwin. *Photo by S. Sanjdorj.*

money on targets that are not characteristic of delivering a major discovery creates setbacks. Perseverance and tenacity are imperative attitudes for the team to make right decisions. The key to success is innovative approach combined with effective cooperation between various disciplines, particularly geologists, geophysicists, and geochemists.

Corporate support is critically imperative for the discovery success. In the case of Oyu Tolgoi, support from the leadership of BHP Discovery until the end of the 1990s played a pivotal role in the dynamic development of the project. All this was expressed both in the allocation of adequate funds for exploration, but also in the provision of required assistance and technical support from the company leadership. Courage in the decision-making process, faith in the success, and a clear vision of the efficient path to discovery characterized the history of the early stage at Oyu Tolgoi. Starting from Magma Copper, and then under BHP, our team approach played a decisive role in the success. The desire to deliver a major mineral discovery was a significant incentive and an effective engine for all participants involved in the early stages of prospecting and exploration.

Discovery of Oyu Tolgoi validated and proved the idea of Paleozoic copper–porphyry deposits still preserved in the old volcanic belts. Paleozoic volcanic belts of the Altaids extending from Central Asia through South Gobi remain prospective for future porphyry copper discoveries, which are waiting for their discoverers. Numerous principle learnings from our success could be applied in other geologic terrains as well and to different types of deposits. We hope these learnings will be useful for future discoverers.

The authors of this book express their appreciation to all those who contributed for the preparation of this book and consequent improvement of its script. We are grateful to all geologists colleagues and support staff who worked with us in the field at all stages of the Oyu Tolgoi early discovery. Our most profound gratitude goes to our families for their dedicated support of our aspirations to produce a significant mineral discovery in the vast stretches of Mongolia. We recognize and acknowledge the companies, under which auspices we worked in the field, and who provided funds and support for our discovery efforts. We are grateful to the publishers, who on the pages of this book made it possible for us to share our observations and our knowledge gathered in the field during the early stage of the Oyu Tolgoi discovery. We are confident that our initial efforts for the major copper discovery in the Mongolian Gobi laid out a foundation for future development of the copper industry in southern Mongolia. Oyu Tolgoi project brought tangible results for the entire mining business of Mongolia and ultimately to the whole country with its hard-working people. Oyu Tolgoi is also a remarkable page in the history of the global copper business and mining industry as a whole.

We acknowledge Dmitri Diakov and Dr. John-Mark Staude for their helpful suggestions and advice for this book. We are especially profoundly grateful to Harry Muntanion for his in-depth critique and very useful recommendations during his careful editing while reviewing the text of this manuscript. Without all their effective assistance, the completion of this manuscript would not be possible.

Appendix 1

Table of Prospects Visited During First Field Campaign

TABLE A1.1 Prospects Visited by Team 1

S. No	Name of Prospects	Mineralization Type	Scale/Mapped Area, km^2	Drilling Line, m/ Number of Drillholes	Trenches (m^3)/Shafts (m)	Sampling (Channel/ Grab Sample)	Geochemistry	Geophysics
1	Bayan Uul	Copper tourmaline veins	1:10,000/100	2,510	11,240	1,360/3,221	Cu, 0.026%; Mo, 0.016%	IP (3%–4.5%)
2	Budagt	Porphyry copper	nd	443.3/3	3 trenches	nd	nd	nd
3	Chandman Uul 3	nd	1:50,000; 1:5,000/1.5	71	Trenches		1:10,000	
4	Chandman Uul 2	nd	1:50,000	No	Trenches		1:10,000	
5	Dzaan Khudag	Porphyry copper?	1:50,000; 1:5,000	~600	20 trenches		1:10,000	nd
6	Morogtsog Khudag	Porphyry copper?	1:50,000	No	Trenches		1:10,000	Magnetics (100 × 20, 44.8 sq. km)
7	Profile 5	Porphyry copper?	1:50,000; 1:25,000; 1:5,000	No	Trenches	nd	1:10,000	
8	Ulaan Khudag 2	Porphyry copper with skarn	1:50,000	73	159	nd	1:10,000	
9	Ulaan Khudag I	Porphyry copper	1:50,000; 1:5,000/4.2	?/7 drl	21 trenches	nd	1:10,000	

10	Jargalant	Porphyry copper?	No	No	No	32	44 samples	
11	Gashuun Khudag	Vein	1:25,000	?/4	2 trenches	Cu–Al	nd	nd
12	Khadat	Porphyry copper	1:50,000/120; 1:1.000/0.4	197.1	9,300/293	100/260	nd	nd
13	Kharmagtai	Breccia pipes						
14	Khatsar	No	nd	nd	7 trenches	No	No	No
15	Khogtsot Uul	Skarn	nd	No	No	No	No	No
16	Zuun Khunguit	No	nd	No	No	No	No	
17	Khunguit	Porphyry copper	1:10,00/12; 1:5,000/2	1,926 (2531.5[a])	5,002 (1772[a])	1,318/901	1:10,000 (7 sq. km)	IP, magnetics
18	Khoshuut	Native copper, zeolites	No	No	Several trenches	No	No	No
19	Khutag Uul	Skarn	1:2,000	No	240/25	107	No	No
20	Khuv Khushaat	Skarn	nd	No	2 trenches	nd	nd	nd
21	Tov and Baruun Huzhur	Porphyry copper?	1:10,000	No	12,438.4	No	Cu, 0.002%–0.03%	nd
22	Mandakh	Copper veins	1:50,000/10	318	300	No	Cu, 0.006%–0.08%	nd

Continued

TABLE A1.1 Prospects Visited by Team 1—cont'd

S. No	Name of Prospects	Mineralization Type	Scale/Mapped Area, km^2	Drilling Line, m/ Number of Drillholes	Trenches (m^3)/Shafts (m)	Sampling (Channel/ Grab Sample)	Geochemistry	Geophysics
23	Mogoin Gol	Secondary quartzites in andesite						
24	Haran	No	1:200,000				Cu, 0.03%–0.6%; Mo, 0.005%–0.01%	
25	Narangiin Khudag	No	1:200,000	No	No	No	No	No
26	Nariin Khudag	Copper tourmaline veins	1:50,000/326	2,119	2,577.6	975/~700	Cu, 0.003%–0.03%	IP (3.0%–4.5%)
27	Oyut Ovoo	Porphyry copper?	1:10,000/200	901.1	1,239/118	279/581	Cu, 0.005% (2.5 sq. km); Mo, 0.0002%; Pb, 0.003%	
28	Oyut Tolgoi	Skarn	1:200,000	No	No	No	No	No
29	Shine Us Khudag	Porphyry copper?	1:200,000	No	No	No	No	No
30	Shuteen	Porphyry copper?	1:50,000/218; 1:10,000/16.3	2,092	5,792.2	472/962	Cu up to 0.05%; Mo up to 0.003%; Zn up to 0.03%; Ag, 1–10 g/t	IP (4%–6%)

31	Teshig	Skarn	1:200,000	nd	Trenches	no	no	no
32	Tsagaan Suvarga	Porphyry copper	1:50,000; 1:10,000	Drilled	Numerous	Regular	Cu, 0.53%; Mo, 0.018%	IP
33	Tsenher Khudag	nd	1:200,000				Cu, 0.1%–1.0%; Bi, 0.003%; Ag, 0.1–3.0 g/t	?
34	Ulaan Tolgoi 1	Cu-porphyry?	1:10,000/50	411	9 trenches	70	Cu, 0.03	IP (3%–8%)
35	Ulaan Tolgoi 2	Cu-porphyry, vein	1:25,000	?/6	2 trenches	nd		abs
36	Olzii Ovoo	Skarn	1:10,000/50	550.9	104	25/260	Cu up to 0.005%; Zn up to 0.007%; Pb up to 0.01%	?

S. No	Prospect Name	Hydrothermal Alterations	Mineral Composition	Dimensions (Length/ Thickness or Width/ Depth, m)	Assays, Cu, %	Assays, Mo, %	By-products
1	Bayan Uul	Propylitic, silica, tourmaline, sericitic, argillic,	pt, cp, rarely mol	2,300–3,000/600/—	0.14–0.32	up to 0.02	Au, 0.22–2 g/t; Ag, 4.3–15 g/t
2	Budagt	Epidote, hematite, silica, hornfels	pt, cp, ga	100–150/20–30/—	0.002–0.5		

Continued

TABLE A1.1 Prospects Visited by Team 1—cont'd

S. No	Prospect Name	Hydrothermal Alterations	Mineral Composition	Dimensions (Length/ Thickness or Width/ Depth, m)	Assays, Cu, %	Assays, Mo, %	By-products
3	Chandman Uul 3	Skarn, silica	pt, cp, sp	300/50/?	up to 1.79	0.001	Au up to 1 g/t, Ag up to 7 g/t, Pb up to 0.3%; Sb up to 0.5%
4	Chandman Uul 2	Silica, potassium feldspar	pt, cp, sp	300–500/50/—	0.02–0.1	0.001	Ag, 4.8 g/t
5	Dzaan Khudag	Potassium feldspar, kaolinite, sericitic, limonite	pt, cp, mal	300–400/20–40/—	0.1–0.8	0.007	
6	Morogtsog Khudag	Silica, hematite, sericitic	pt, cp				
7	Profile 5	Silica, potassium feldspar, epidote, chlorite	nd	150–600/80–100/—	0.3–1.11	0.007	
8	Ulaan Khudag 2	Feldspar, epidote, sericitic, chlorite	nd	300/5–10/—	0.1–0.3	0.003	
9	Ulaan Khudag I	Potassium feldspar, kaolinite, sericitic	nd	700–900/15–25/—	0.3–0.7	0.005	

10	Jargalant	Silica, chloritic, propylitic	mg, pt, cp, bn, mal, az	22,000/40–80/—	0.1-5.0		Ag, 5–30 g/t; Au up to 0.3 g/t
11	Gashuun Khudag	Epidote, potassium feldspar, chlorite	pt, cp, ga, sp, mal	100–300/0.1–1/200	0.03–1.0		
12	Khadat	Silica, hornfels, epidote	Mal, az, tn, rarely cp, pt, ap, sp, hm	150/10–30/100	0.1–0.5		
13	Kharmagtai	Silica, propylitic	pt, cp				
14	Khatsar	Potassium feldspar, chlorite	cp, mal, ga	10–300/0.3–3/—	0.004–0.5		
15	Khogtsot Uul	Skarn	cp	70–160/30/—			
16	Zuun Khunguit	?		430–550/20–90/—			
17	Khunguit	Silica, potassium feldspar, epidote, tourmaline	mg, mol, cp, pt	N 1,000/300/200–320; W 300–650/100	0.1–0.3		
18	Khoshuut	Epidote		150/m/—	3.0–10.0?		Ag, 6 g/t
19	Khutag Uul	Silica (intense), skarn, hornfels	pt, cp, mal, az	800/10–240/—			
20	Khuv Khushaat	Skarn	mal, az	200/4/—	0.05–1.25		

Continued

TABLE A1.1 Prospects Visited by Team 1—cont'd

S. No	Prospect Name	Hydrothermal Alterations	Mineral Composition	Dimensions (Length/Thickness or Width/Depth, m)	Assays, Cu, %	Assays, Mo, %	By-products
21	Tov and Baruun Huzhur	Silica, tourmaline	mal, az, mg, hm, pt, rare cp, bn, ga	1,200–2,000/300–700/—	up to 0.9		Pb, Zn up to 0.9%, Ag up to 60–300 g/t
22	Mandakh	Silica, tourmaline	cp, bn, mal	110–150/10/—	0.1–1.16		
23	Mogoin Gol	?					
24	Haran	?					
25	Narangiin Khudag	Hornfels	mal, cp	100/50/—	0.05		
26	Nariin Khudag	Tourmaline, potassium feldspar, silica	cp, pt, bn, mal, az	100–400/5–23/180	0.32–0.65		Au, 1 g/t; Ag, 5 g/t
27	Oyut Ovoo	Potassium feldspar	mal, tn, az, chr	100/25/10–150	0.01–0.15		Ag, 0.5 g/t; Pb, 0.007%–0.015%
28	Oyu Tolgoi	Skarn, silica, chlorite	cp, bn, pt, pr	70/5–8/—	0.02-2.0		

29	Shine Us Khudag	Hornfels	mal, cp	150/50/—	0.03–0.08		
30	Shuteen	Advanced argillic, propylitic, intense silica, alunite, sericitic, tourmaline	cp, mal, native gold	7,000/1,500–2,000/ 190–280	0.01–0.5		
31	Teshig	Skarn, hornfels	pt, cp, bn, native gold, mal, az	40–100/0.5–6/0–100	0.1–3.0		Au up to 30 g/t
32	Tsagaan Suvarga	Intensive silica, sericite, propylitic	pt, cp, mal		0.55	0.018	Au, Ag
33	Tsenher Khudag	Silica					
34	Ulaan Tolgoi 1	Potassic feldspar, rarely sericite	mal, az, pt, cp	200/4–50/—	0.1–0.3		
35	Ulaan Tolgoi 2	Silica, sericitic, greisen, tourmaline, potassium feldspar	mal, az, cp, mol	220/12–15/—	0.2–1.0 (oxidized), 0.015–0.3		Ag, 1–15 g/t
36	Olzii Ovoo	Skarn	sp, rarely cp	400/200/100–150	0.07		Zn, 0.5%

abs, absent; *ap*, apatite; *ar*, arsenopyrite; *az*, azurite; *bn*, bornite; *chp*, chrysoprase; *chr*, chrysocolla; *cp*, chalcopyrite; *cu*, cuprite; *cv*, covellite; *drl*, drillholes; *ga*, galena; *hm*, hematite; *IP*, induced polarization; *mal*, malachite; *mg*, magnetite; *mol*, molybdenite; *nd*, no data; *No*, not established; *pr*, pyrrhotite; *py*, pyrite; *recon*, reconnaissance; *sp*, sphalerite; *tn*, tenorite.
[a]*Various sources.*

Appendix 2

Table of Prospects Visited by Team 2 During First Field Campaign

TABLE A2.1 List of Prospects Visited by Team 2

S. No	Name of Prospects	Mineralization Type	Scale/ Mapped Area, km^2	Drilling Line, m/Number of Drillholes	Trenches (m^3)/ Shafts (m)	Sampling (Channel/ Grab Sample)	Geochemistry	Geophysics
1	Airag Uul	VMS	1:200,000	no	no	nd	nd	no
2	Altan Uul (Khanan)		1:200,000	no	3 trenches	nd	no	no
3	Altan	Skarn	1:200,000; 1:50,000	—/1	3 trenches	—/13	nd	no
4	Altan Khad		1:200,000	no	no	18	no	no
5	Baidrag (Kheseg No. 29)		Recon	—/2	2 trenches	nd	100 × 20	no
6	Bayan Tsagaan		1:200,000/ 110; 1:10,000/26	25.8/1	720	300/—	100 × 20 (8,500 samples)	no
7	Borts Uul	VMS	1:200,000; 1:50,000; 1:1,000	549/5	6,978/76, trenches	337/92	100 × 20	no
8	Olongiin Suuj Khudag	Stratabound	1:200,000/ 110; 1:10,000/9	165/2	1,978	375/56	100 × 20 (7 profiles, 360 samples)	no
9	Jargalant		Recon/ 70 km	134/1	3,025	397/34	168 samples	

10	Zost Uul 1	Skarn	1:200,000	no	3 trenches	nd	no	no
11	Zost Uul 2							
12	Khondloi	VMS	1:200,000; 1:50,000	~500/10	927	nd	500 × 100	no
13	Khalzan Bulag 1	Porphyry copper	1:200,000	—/3	4 trenches	nd	no	no
14	Khalzan Bulag 2 and 3	Porphyry copper	1:200,000	no	no	12	no	no
15	Khar Uul	VMS	1:200,000	no	2 trenches	9/—	no	no
16	Hariin Khar Uul (Khuren Tolgoi)	VMS	1:200,000	no	1 trench	nd	no	no
17	Khukh Bulgiin Khundii	Skarn	1:200,000; 1:50,000	no	2 trenches	no	no	Magnetics
18	Khukh Adar	VMS	1:10,000/10	—/9	19 trenches	nd	nd	no
19	Khuren Tolgoi (Bayan Bulag)	VMS	1:50,000/ 34.3 km^2	118.5/2	535	55/5	400	no
20	Mongoin Shand	Skarn	1:200,000	no	no	16	no	no
21	Tamiriin Gol		1:25,000/ 63.5 km^2	1,203.5/9	23,298	352/218	50 × 500	Magnetics (250 × 25)

Continued

TABLE A2.1 List of Prospects Visited by Team 2—cont'd

S. No	Name of Prospects	Mineralization Type	Scale/ Mapped Area, km^2	Drilling Line, m/Number of Drillholes	Trenches (m^3)/ Shafts (m)	Sampling (Channel/ Grab Sample)	Geochemistry	Geophysics
22	Prospect no. 28		1:50,000	2	2 trenches	12	no	no
23	Prospect no. 31		1:200,000	no	2 trenches	19	no	no
24	Prospect no. 34		1:200,000	no	no	34	no	no
25	Prospect no. 35		1:200,000	no	no	2	no	no
26	Prospect no. 62		1:200,000	no	no	7	no	no
27	Prospect no. 68							
28	Omnod	Skarn	1:200,000	no	trench	28	no	no
29	Zamiin Khadjuu no. 67	Stratabound and skarn	1:200,000	no	trench	nd	no	no
30	Saran Uul	Porphyry copper	Prospecting evaluation	—/14	9 trenches	nd	nd	no

31	Shar Tal		1:200,000	no	5 trenches	no	no	no
32	Shirt	Skarn	1:200,000	—/1	? trenches	nd	no	no
33	Shuvuun khar Uul	VMS	1:200,000; 1:50,000	—/1	nd	22/17	no	no
34	Tsanhir Del (#56)	VMS	1:200,000	no	no	—/4	no	no
35	Ulaan Khajuu							
36	Omnogobi	Stratabound	1:50,000/14	104/1	4 trenches	nd	500 × 50 (346 traverses)	no
37	Zambain Khudag	Skarn	1:200,000	55/2	180	nd	1:200,000	no

S. No	Prospect Name	Hydrothermal Alterations	Mineral Composition	Dimensions (Length/ Thickness or Width/ Depth, m)	Assays, Cu, %	Assays, Mo, %	By-products
1	Airag Uul	Silica, carbonate, hematite, jarosite	pt, mal	2,000–3,000/5–100/ —	0.01–1		Au, 0.2 g/t; Ag, 1.5 –15 g/t
2	Altan Uul (Khanan)	Quartz, sericite, epidote	400–500/5–10/—				
3	Altan	Skarn	bn, cp, mal, az	Up to 200/5/—	0.85–2.3		Au, 1.8 g/t
4	Altan Khad	Potassium feldspar	cp	2,500/3–4/—	0.01–0.3		
5	Baidrag (Kheseg no. 29)	Silica	mal, az, pt, cp	Small	0.3–0.35		

Continued

TABLE A2.1 List of Prospects Visited by Team 2—cont'd

S. No	Prospect Name	Hydrothermal Alterations	Mineral Composition	Dimensions (Length/Thickness or Width/Depth, m)	Assays, Cu, %	Assays, Mo, %	By-products
6	Bayan Tsagaan	Quartz, serpentinites	pt, cp	up to 150/1–30/—	0.1–0.22		
7	Borts Uul	Chlorite, epidote	cp, cc, bn, cu, cv	150–500/5–30/—	0.33–1.31		Ag up to 5 g/t
8	Olongiin Suuj Khudag		mal, az	100/0.2–0.6/—	0.1–1		
9	Jargalant	Quartz, carbonate	mal, az, cp	Small	0.01–0.6		Ag, 5 g/t
10	Zost Uul 1	Hornfels, skarn, silica	cp, mal	2,500/5–15/—	0.02–2.16		Au up to 1 g/t
11	Zost Uul 2						
12	Khondloi			6,000/50–200/—	0.15–8		1–5 (Zn), 0.7–1.5 (Pb)
13	Khalzan Bulag 1	Silica, sericite	mal, chr	500–700/10–120/100	0.05–5	0.002–0.01	
14	Khalzan Bulag 2 and 3	Hornfels, skarn, epidote	cp, pt, mal	200/4–10/—	0.2–3.34		Au
15	Khar Uul	Intense silica	cp, mal	500/150/—			
16	Hariin Khar Uul (Khuren Tolgoi)	Skarn	cp, bn, native gold	40–160/15–20/—			Au, 2 g/t
17	Khukh Bulgiin Khundii	Silica	pt	1,300/150/—	0.01–2	0.002	Ag, 0.5–15 g/t

18	Khukh Adar	Silica, argillic, graphite, sericitic, limonite	cp, pt	1,000–2,000/30–140/ 100–150	0.05–15.4		Ag up to 20–70 g/t
19	Khuren Tolgoi (Bayan Bulag)	Silica, hematite, epidote	Cp, bn, mal, a3, cv	900–1,000/20–80/—	0.1–2.04		Cu, Ag
20	Mongoin Shand	Silica, potassium feldspar, carbonate, skarn	cp, hm	1,000/1.5–20/—	0.02–15	Up to 0.03	Ag up to 100 g/t
21	Namiriin Gol	Argillic, silica, greisen	cp, pt, sp	25–350/7–15/—	0.02–6.28		Ag, 1–79 g/t
22	Prospect no. 28				???		
23	Prospect no. 31	Argillic, silica, limonite	cp, mal	Limited six	0.1–2		Zn, Bi
24	Prospect no. 34	Silica	cp, pt, mal	100/1–10/—	0.1–6		Ag up to 50 g/t
25	Prospect no. 35	Silica		700/—/—	0.05–3		
26	Prospect no. 62	Quartz		Limited size	0.5–1		Ag
27	Prospect no. 68	Silica	cp, mal	100–30/2–500/—	0.01–0.6		
28	Omnod	Skarn	cp	1,000/2–5/—	0.01–10		Ag up to 30 g/t
29	Zamiin khaduuj no. 67	Skarn, silica, chlorite	mal, az, bn, cp	10–700/1–50/—	0.01–2.26		Au, 0.8 g/t; Ag, 5–100

Continued

TABLE A2.1 List of Prospects Visited by Team 2—cont'd

S. No	Prospect Name	Hydrothermal Alterations	Mineral Composition	Dimensions (Length/ Thickness or Width/ Depth, m)	Assays, Cu, %	Assays, Mo, %	By-products
30	Saran Uul	Potassium feldspar, albite, silica, biotite, chlorite, kaolinite	cp, pt, mal	1,400–900/200–600/ 300	0.37	0.001–0.037	Au up to 0.1 g/t
31	Shar Tal		pt	3,000/5–90/—	0.01–1		Zn
32	Shirt	Silica, hematite		50–130/0.5–5/—	0.3–17.78	Up to 0.01	Ag, 10 g/t; Au, 0.8 g/t, Zn, Pb
33	Shuvuun Khar Uul	Quartz, sericitic, carbonate	pt, mal	100–150/40–50/—	0.2–0.8	0.02	Au, 0.05–2 g/t
34	Tsanhir Del (#56)	Silica	pt, cp, mal	400–500/20–100/—	0.2–1		
35	Ulaan Khajuu				???		
36	Omnogobi		cp, mal, az	1,500/50/—	0.13–1.74		
37	Zambain Khudag	Hornfels, skarn, epidote, potassium feldspar	cp, mal, az	400–600/5–10/—	0.02–0.5		Au, 0.01–3 g/t

az, azurite; *bn*, bornite; *cc*, chalcocite; *chr*, chrysocolla; *cp*, chalcopyrite; *cu*, cuprite; *cv*, covellite; *hm*, hematite; *mal*, malachite; *nd*, no data; *no*, not established; *sp*, sphalerite; *pt*, pyrite.

Appendix 3

Historic Milestones and Chronology of BHP's Work at Oyu Tolgoi

1995	April	Signed the Erdenet–BHP Agreement with 50/50 ownership to conduct joint geological exploration for copper in Mongolia.
	June	Started reconnaissance programs with two field teams.
	August	Visited 73 copper prospects and mineralization points. The work focused on obvious manifestations of copper on the surface. Upon completion of the fieldwork, decision was to refocus the program on search of superficial zones of leaching.
	September	Joint venture meeting in Tucson. JV is encouraged to continue its search.
	December	BHP is buying Magma Copper; all exploration programs become part of the global geological unit of BHP Minerals.
1996	June	The decision to close the Joint Venture Erdenet–BHP. BHP's decision to continue prospecting and exploration through its 100% owned entity in Mongolia.
	July	BHP Minerals filed an application for registration of the representative office.
	August	BHP Minerals registered its representative office in Mongolia and received permission to continue reconnaissance.
	September	Reconnaissance resumed. BHP field team found an outcrop of Central Oyu Tolgoi.
	October	BHP Minerals applied for an exploration license over Oyu Tolgoi and several other prospects of interest.
1997	January	BHP registered a branch in Mongolia. Exploration license for an area of 1,350 km^2 at Oyu Tolgoi was issued to BHP Branch.
	April	Fieldwork began on Oyu Tolgoi (topographic survey, geophysics, geological mapping, and geochemistry).

	June	Completed geophysical (magnetic survey and IP) and continued mapping and geochemistry.
	September	Launched the first phase of drilling at Oyu Tolgoi.
	October	First six drillholes drilled. Borehole OT3 crossed chalcocite mineralization in the Central Oyu Tolgoi; drillhole OT4 intercepted stockwork with quartz–magnetite–bornite mineralization on South Oyu Tolgoi.
	December	The results of analysis of samples from the drilled drillholes confirmed the presence of chalcocite mineralization at Central Oyu Tolgoi and hypogene copper–gold mineralization at South Oyu Tolgoi.
1998	April	Fieldwork at Oyu Tolgoi resumed. TEM geophysical survey within Central Oyu Tolgoi. Additional IP survey held on the South and West Oyu Tolgoi.
	May	The second phase of drilling started. Completion of the geochemical survey at Oyu Tolgoi hills.
	September	Conducting magnetic airborne survey and beginning of the third phase of drilling.
1999	January	Data analysis, resources, practical advice.
	February	The expert porphyry group of BHP concluded that Oyu Tolgoi deposit meets its requirements for tier 2 deposits and fell short of reaching tier 1 status. Decision made to find investors for the project.
	March	Samples and brief information about Oyu Tolgoi presented at the PDAC Conference in Toronto, Canada. Core shack section demonstrated the drill core from Oyu Tolgoi.
	May–September	Search of investors for the project. Western Mining expressed interest but did not receive corporate approval.
	October	Ivanhoe Mines expressed interest in the project and its intention to discuss the terms of the agreement.
2000	May	Agreement between BHP Minerals International, Inc. and Ivanhoe Mines reached and signed on May 5, 2000.
2001	March	Reduction of the licensed area at Oyu Tolgoi. BHP and Billiton agreed to merge their companies into one BHP Billiton diversified mining company.
2002	April	BHP Billiton received notification that Ivanhoe Mines fulfilled the condition of the expenditure obligation at Oyu Tolgoi according to the agreement.
	June	Confirmation that the mineralization delineated by Ivanhoe Mines through their exploration drilling at Oyu Tolgoi did not reach the thresholds established by the agreement for BHP Billiton return to the project.
	July	BHP Billiton transfers the licenses for Oyu Tolgoi Ivanhoe Mines.
	September	Hugo Dummett tragically died in a car accident in South Africa.
2003	November	Ivanhoe Mines acquires 2% royalty on Oyu Tolgoi from BHP Billiton and becomes full owner of the project.
	September	Far North Oyu Tolgoi was renamed after Hugo as Hugo Dummett deposit. The mine to be built will also bear his name as Hugo Mine.

2004	October	Rio Tinto enters into the agreement with Entrée Gold.
2005	March	Robert Friedland's speech in Tampa Florida.
	June	Rio Tinto increases its stake in Entrée Gold to 11%.
	August	BHP Billiton reviews the Oyu Tolgoi project. Recommendation is made to the board to invest in the project but failed to receive a full support.
2006	October	Rio Tinto signs agreement with Ivanhoe Mines for a step-by-step purchase of interest in the Oyu Tolgoi through project funding in exchange of financing the project to build the mine
2008	March	Ivanhoe announces the resource of Heruga Deposit copper-gold deposit
2009	October	Rio Tinto and Ivanhoe Mines signed long term investment agreement with the Government of Mongolia
2012	January	Rio Tinto reached 51% ownership of Oyu Tolgoi
2013	January	Production of the first concentrate from Oyu Tolgoi
	July	First shipment of the concentrate from Oyu Tolgoi

Appendix 4

Explanation of Types of Veinlets in Porphyry Systems

For better interpretation, we present nomenclature of veinlets in stockworks of copper porphyry systems published by Gustafson and Hunt (1995). Richard Sillitoe (2010) also illustrates the variety of typical veinlets in stockworks.

Veins of type "A" are usually the earliest among diverse veinlets in stockwork system. They often have quartz, quartz–feldspar composition with disseminated pyrite, chalcopyrite, and sometimes molybdenite. Quartz is uniformly grainy. Salvages often develop feldspar. Thickness of type A veinlets vary from millimeters to 2.5 cm and more. The length proportions to meters and can reach several meters. Very often, they are curvilinear, apparently formed when the intrusion was still warm to full solid state.

Veins of type "B" formed after the formation of the veins A and always cross the former. Thickness ranges from 0.5 to 5.0 cm. Length ranges from a few meters to tens of meters. Sulfide mineralization is present in the form of strips, following parallel contacts with veinlets with rocks, and is composed of pyrite–chalcopyrite, sometimes with the presence of molybdenite. Changes along salvages are usually with quartz–feldspar.

Veins of the type "D" later cut through the veins of type "A" and "B." Salvages are often accompanied by quartz–kaolin–sericite halo. More often than not are they primarily composed of quartz with sulfides pyrite–chalcopyrite, sometimes with inclusions of enargite, sphalerite, galena, anhydride, and so forth from millimeters up to 7.5 cm. The length varies from a few meters to several tens of meters.

In addition, researchers use various additional nomenclatures for various veinlets encountered in the porphyry systems, such as EH or EB, which indicate early hornblende or biotite veinlets. All the abovementioned three types of (A, B, and D) veins are basic and most commonly used when describing veinlets in the porphyry stockworks.

Bibliography

ADEQ, 2013. Aquifer Protection Plan Fact Sheet. BHP Billiton San Manuel Operation. Arizona Department of Environmental Quality, p. 10. Aquifer Protection Permit 100421.
Berzina, A.P., Sotnikov, V.I., 1991. Ore-magmatic systems in various geodynamic environments (Mongolian copper-molybdenum deposits). Geologiya i Geofisika 9, 91–97 (in Russian).
Briggs, D., 2013. History of the San Manuel-Kalamazoo Mine, Pinal County, Arizona, pp. 1–8.
Burenhuu, E., Gotovsuren, A., Badarch, G., Davastereen, A., 1995. Report of complex geological study at Galba Gobi area (Quadrangle K48-XI), scale 1:200,000. Ministry of Geology file number 4877 DII.
Chovan, M., Gregush, J., Moravek, R., 1983. Geology of Shuteen district. Library of Ministry of Geology. Geological report 3615.
Chovan, M., Gregush, J., Moravek, R., Delgertsogt, B., Lukach, F., 1984. Report about results of reconnaissance work on volcano-plutonic structures at the Ih Shah, Harmagtai, and Shuteen areas of South Gobi Aimag in 1983. Team N35 Vol. 1. Ministry of Geology file number 3696.
Cox, D.P., Singer, D.A., 1986. Mineral Deposit Models, 1693. USGS, Bulletin, 379 p.
Danielsen, V., Jan 6, 2009. The northern miner 1998 "mining man of the year" hugo dummett. Northern Miner.
Force, E.R., Dickinson, W.R., Hangstrum, J.T., 1965. Tilting history of the san manuel-kalamazoo porphyry system, Southeastern Arizona. Economic Geology 90, 67–80.
Forberger, V., 2002. Magma Copper Case Study, Center for Urban and Regional Policy. Northeastern University, pp. 1–49.
Garamjav, D., 2014. In: Batbilig, S. (Ed.), Dursamj-1. Ulaanbaatar, DCI, Havleliyn company, p. 543 (in Mongolian).
Gavrilova, S.P., Maksimyuk, I.E., 1990. Phases of formation of the Eredentuin-Obo copper-molybdenum porphyry deposit (Mongolia). In: Geology of Ore Deposits (Geologiya Rudnykh Mestorozhdeniy), vol. 32, pp. 3–17, 6, (in Russian).
Geology of Mongolian People Republic. In: Marinov, N.A., Khasin, R.A., Hurts, Ch (Eds.), 1977. Mineral Resources, vol. 3. Nedra, Moscow, p. 703 (in Russian).
Gerel, O., 1990. Petrology, geochemistry and mineralization of sub alkaline Mesozoic magmatism in Mongolia. Dr. of Science thesis. Irkutsk, p. 40 (in Russian).
Gerel, O., Munkhtsengel, B., 2005. Erdenetiin Ovoo porphyry copper-molybdenum deposit in northern Mongolia. In: Seltman, R., Gerel, O., Kirwin, D.J. (Eds.), Geodynamics and Metallogeny of Mongolia with Special Emphasis on Copper and Gold Deposits: SEG-IAGOD Field Trip, CERCAMS. Natural History Museum, London, pp. 85–103.
Gustafson, L.B., Hunt, J.P., 1975. The porphyry copper deposit at El Salvador, Chile. Economic Geology 70, 857–912.
International Directory, 1993. International Directory of Company Histories, vol. 7. St. James Press.

International Directory of Company Histories, January 28, 2018. Magma Copper Company. Encyclopedia.com. Retrieved from: http://www.encyclopedia.com.

Khasin, R.A., 1971. Main Features of Evolution of Magmatism in Mongolia. In: Magmatism and Metallogeny of Mongolian People Republic. Nauka, Moscow, pp. 7–40 (in Russian).

Khasin, R.A., Marinov, N.A., Khurtz, Ch., Yakimov, L.I., 1977. The copper-molybdenum deposit at Erdenetiin Ovoo in northern Mongolia. Geology of ore deposits. No. 6, pp. 3–15 (in Russian).

Koval, P.V., Gerel, O., Smirnov, V.N., Tseden, Ts, 1985. Porphyritic intrusion from the Erdenet ore deposit: petrography and chemistry; Problems of Geology and metallogeny of Mongolia. Ulaanbaatar, pp. 67–103 (in Russian).

Koval, P.V., Gotovsuren, A., Ariunbileg, S., Lubatorov, Y.I., 1989. On prospecting for porphyry copper mineralization in intracontinental mobile zones (Mongol-Okhotsk belt, Mongolian People's Republic). Journal of Geochemical Exploration, 32, 369–380.

Koval, P.V., Gerel, O., 1986. Volcanic association of porphyry copper mineralization in the Mongol-Okhotsk intraplate mobile zone. In: Geochemistry of volcanics from different geodynamic settings. Nauka, Novosibirsk, pp. 9–93 (in Russian).

Kovalenko, V.I., Tsaryeva, G.M., Goreglyad, A.V., Yar, V.V., Troitsky, V.A., Hervig, R.L., Farmer, G.L., 1995. The peralkaline granite-related Khaldzan -Buregtey rare metal (Zr, Nb, REE) deposit. western Mon Economic Geology 90, 530–547.

Kovalenko, V.I., Tsaryeva, G.M., Goreglyad, A.V., Yar, V.V., Troitsky, V.A., Hervig, R.L., Farmer, G.L., 1995. The peralkaline granite-related Khaldzan–Buregtey rare metal (Zr, Nb, REE) deposit, western Mongolia. Economic Geology 90, 530–547.

Kovalenko, V.I., Yarmoluyk, V.V., Sal'nikova, E.B., Kozlovsky, A.M., Kotov, A.B., Kovach, V.P., Savatenkov, V.M., Vladykin, N.V., Ponomarchuk, V.A., 2006. Geology, geochronology, and geodynamics of the Khan Bogd alkali granite pluton in southern Mongolia. Geotectonics 40, 450–466.

Krajick, K., 2001. Barren Lands: an Epic Search for Diamonds in the North America Arctic. Harry Holt and Company LLC, p. 442.

Lamb, M.A., Cox, D.P., 1998. New 40Ar/39Ar age data and implications for porphyry copper deposits in Mongolia. Economic Geology 93, 524–529.

Lowell, J.D., 1968. Geology of the kalamazoo orebody, San Manuel district, Arizona. Economic Geology 63, 645–654.

Lowell, J.D., Guilbert, J.M., 1970. Lateral and vertical alteration-mineralization zoning in porphyry ore deposits. Economic Geology 65, 373–408.

Manske, S., Paul, A., 2002. Geology of a major new porphyry copper center in the superior (pioneer) district, Arizona. Economic Geology 97, 197–220.

Ochirbat, P., 2010. Oyu Tolgoi: past, present and future. ADMON, p. 281 (in Mongolian).

Paul, A.H., Knight, M.J., 1995. Replacement ores in the magma mines, superior, Arizona. In: Pierce, F.W., Bolm, J.G. (Eds.), Porphyry Copper Deposits of the American Cordillera, vol. 20. Arizona Geological Society Digest, pp. 366–372.

Paul, A.H., Manske, S.L., 2005. History of exploration at the magma mine, superior, Arizona. In: Rhoden, H.N., Steininger, R.C., Vikre, P.G. (Eds.), Geological Society of Nevada Symposium 2005, Window to the World, Reno, Nevada, May 2005, Proceedings, vol. 1, pp. 629–638.

Pebble, 2014. Resource. Northern Dynasty Minerals Ltd @ http://www.northerndynastyminerals.com/ndm/Pebble_Drill_Program.asp.

Perelló, J., Cox, D., Garamjav, D., Sanjdorj, S., Diakov, O.S., Schissel, D., Munkhbat, T.-O., Gonchig, O., 2001. Oyu Tolgoi, Mongolia: siluro-Devonian Porphyry Cu-Au-(Mo) and high sulfidation Cu mineralization with a Cretaceous chalcocite blanket. Economic Geology 96, 1407–1438.

Porter, T.M. (Ed.), 2005. Super Porphyry and Gold deposits: A Global Perspective. PGC Publishing, Adelaide, Australia v.2. pp. 525–543.

Resource Investor, 2005. Nothing like it on Planet Earth – Robert Friedland's Tour D'Tolgoi. Reprints from speech delivered in Tampa, Florida on March 7, 2005. http://www.resourceinvestor.com/2005/03/06/nothing-it-planet-earth-robert-friedlands-tour-d-tolgoi.

Sandbak, L.A., Alexander, G.H., 1995. Geology and rock mechanics of the Kalamazoo orebody. In: Pierce, F.W., Bolm, J.G. (Eds.), Porphyry copper deposits of the American Cordillera: Tucson, 20. Arizona Geological Society Digest, San Manuel, Arizona, pp. 396–423.

Sillitoe, R.H., 2010. Porphyry copper systems. Economic Geology 105, 3–41.

Sotnikov, V.I., Berzina, A.P., 1984. Copper-molybdenum mineralization and ore-bearing magmatism in Mongolia: Quadrennial IAGOD Symposium, 6th. September 1984, Proceedings, v. 6, no. 1, Tbilisi, USSR, pp. 445–448.

Sotnikov, V.I., Berzina, A.P., 1989. Prolonged discrete oriented development of ore-magmatic systems in copper-molybdenum formation. Geology and Geophysics (1), 41–45 (in Russian).

Sotnikov, V.I., Berzina, A.N., Economou-Eliopoulos, M., Eliopoulos, D., 2001. Palladium, platinum and gold distribution in porphyry Cu-Mo deposits of Russia and Mongolia. Ore Geology Reviews vol. 18, pp. 95–111.

Sotnikov, V.I., Berzina, A.N., Jamsran, M., Garamjav, D., Bold, D., 1985. Copper deposits of Mongolia: Transactions of the Joint Soviet-Mongolian Scientific-Research Expedition, Nauka, Novosibirsk, vol. 43, p. 223 (in Russian).

The Resolution Copper Project, 2015. http://resolutioncopper.com/the-project/.

Thompson, P., Macklin, R., 2009. The Big Fella: The Rise and Rise of BHP Billiton. William Heineman, Australia, p. 518.

Titley, S.R., 1994. Silver Bell porphyry copper deposit, Silver Bell Mountains, Pima County, Arizona. In: Thorman, Lane, D.E. (Eds.), USGS research on mineral resources, 1994, Part B. Guidebook for field trips. US Geological Survey Circular 1103-B, pp. 77–88. Also available at http://pubs.usgs.gov/circ/1994/1103b/report.pdf.

Turquoise Hill Press Release, August 18, 2004a. Southern Oyu Tolgoi Measured and Indicated Copper and Gold Resources Increased by 109% in New Independent Resource Estimate. Retrieved from: http://www.turquoisehill.com/s/news_releases.asp?ReportID=375402.

Turquoise Hill Press Release, November 10, 2004b. Ivanhoe Mines and Entrée Gold Finalize Earn-in Agreement. Retrieved from: http://www.turquoisehill.com/s/news_releases.asp?ReportID=375411.

Turquoise Hill Press Release, November 18, 2004c. Ivanhoe's Hugo North Deposit Now Believed to be the World's Highest-Grade Copper Porphyry Discovery. Retrieved from: http://www.turquoisehill.com/s/news_releases.asp?ReportID=375414.

Turquoise Hill Press Release, December 6, 2004d. Ivanhoe Mines Extends Hugo North Geophysical Anomaly Four Kilometres North. Retrieved from: http://www.turquoisehill.com/s/news_releases.asp?ReportID=375418.

Turquoise Hill Press Release, February 1, 2006a. New Mineral Resource Estimate Adds 281 Million Tonnes of High-Grade Inferred Resources to the Hugo North Copper-Gold Deposit in Mongolia. Retrieved from: http://www.turquoisehill.com/s/news_releases.asp?ReportID=375479.

Turquoise Hill Press Release, October 18, 2006b. Ivanhoe Mines and Rio Tinto Form Strategic Partnership to Develop Mongolian Copper-Gold Resources. Retrieved from: http://www.turquoisehill.com/s/news_releases.asp?ReportID=375530.

Turquoise Hill Press Release, March 12, 2008. Ivanhoe Mines Reports an Initial 760 Million Tonne Inferred Resource Estimate for the Heruga Copper, Gold and Molybdenum Deposit in Mongolia. Retrieved from: http://www.turquoisehill.com/s/news_releases.asp?ReportID=375610.

Turquoise Hill Press Release, October 9, 2009. Ivanhoe Mines and Rio Tinto Sign Long-Term Investment Agreement with Mongolia to Build and Operate Oyu Tolgoi Copper-gold Mining Complex. Retrieved from: http://www.turquoisehill.com/s/news_releases.asp?ReportID=375664.

Turquoise Hill Press Release, May 11, 2010a. Ivanhoe Mines Releases New Integrated Development Plan for Oyu Tolgoi Copper-gold Mining Complex in Mongolia. Retrieved from: http://www.turquoisehill.com/s/news_releases.asp?ReportID=398902.

Turquoise Hill Press Release, September 8, 2010b. New Drill Hole at Oyu Tolgoi Intercepts 938 Meters of Near-continuous Copper/gold Mineralization between Heruga Deposit and Southern Oyu Deposits. Retrieved from: http://www.turquoisehill.com/s/news_releases.asp?ReportID=420912.

Turquoise Hill Press Release, July 8, 2013. Oyu Tolgoi Begins Concentrate Shipments. Retrieved from: http://www.turquoisehill.com/s/news_releases.asp?ReportID=591364.

Turquoise Hill Press Release, October 16, 2017. Turquoise Hill Announces Third Quarter 2017 Production; Operational Records Achieved during Quarter. Retrieved from: http://www.turquoisehill.com/s/news_releases.asp?ReportID=805251.

Turquoise Hill, February 2017. BMO Global Mining Presentation. http://www.turquoisehill.com/i/pdf/ppt/Turquoise-Hill-BMO-presentation-February-2018-21-Feb-20.pdf.

Wikipedia, 2015a. Retrieved from: https://en.m.wikipedia.org/wiki/Mongolia.

Wikipedia, 2015b. Retrieved from: http://en.wikipedia.org/wiki/Erdenet_Mining_Corporation.

Wikipedia, 2015c. Retrieved from: http://en.wikipedia.org/wiki/San_Manuel_Copper_Minepedia.

FURTHER READING

Atlas of Mineral Resource of the ESCAP Region, 1999. Geology and mineral resources of Mongolia. United Nations 14, 192.

Ballantyne, G., Marsh, T., Hehnke, C., Andrews, D., Eichenlaub, A., Krahulec, K., 2003. The Resolution Copper Deposit, a Deep, High-Grade Porphyry Copper Deposit and the Superior District. SEG Student Chapter, Arizona, pp. 1−12.

Bat-Erdene, K., Rye, R., Hedenquist, J., Kavalieris, I., 2006. Geology and reconnaissance stable isotope study of the Oyu Tolgoi porphyry Cu-Au system, South Gobi, Mongolia. Economic Geology 101, 503−522.

Bat-Erdene, K., Rye, R., Kavalieris, I., Hayashi, K., 2009. The sericitic to advanced argillic transition: stable isotope and mineralogical characteristics from the hugo dummett porphyry Cu-Au deposit, Oyu Tolgoi district, Mongolia. Economic Geology 104, 1087−1110.

Berzina, A.N., Sotnikov, V.I., Polomarchuk, V.A., Berzina, A.P., Kiseleva, V.Y., 1999. Temporal periods of formation of Cu-Mo porphyry deposits, Siberia and Mongolia. In: Stanley, C.J., et al. (Eds.), Mineral Deposits: Processes to Processing, vol. 1. In: Proceedings of the Fifth Biennial SGA Meeting, Balkema, Rotterdam, p. 321−324.

Berzina, A.I., Sotnikov, V.I., Economou-Eliopoulos, M., Eliopoulos, D.G., 2001. Precious metal contents of porphyry Cu-Mo deposits of Russia and Mongolia. In: Piestrzynski, A., others (Eds.), Mineral Deposits at the Beginning of the 21st Century. A.A. Balkema Publishers, Lisse, pp. 699−702.

Berzina, A.I., Sotnikov, V.I., Economou-Eliopoulos, M., Eliopoulos, D.G., 2005. Distribution of rhenium in molybdenite from porphyry Cu-Mo and Mo-Cu deposits of Russia (Siberia) and Mongolia. Ore Geology Reviews 26, 91–113.

Cox, D.P., Diatchkov, S.A., Garamjav, D., Munkhbat, T., Oyun, G., Perello, J., Sanjdorj, S., Carter, S.G., Smail, R.F., 1997. Results of exploration activity in 1997 at Oyu Tolgoi and Ih Shanh areas, South Gobi, Mongolia. Unpublished Internal Report BHP Minerals Exploration, 101.

Erdenet Mining Corp, 2002. 2002-EMC Annual Report.

Gavrilova, S.P., Maksimyuk, I.E., Orolmaa, D., 1984. Features of magmatism and composition of ore of the copper-molybdenum deposit at Erdenetiin Ovoo. In: Endogenic Ore Formations of Mongolia. Nauka, pp. 101–105 (in Russian).

Goldenberg, V.I., Sanjaadorj, D., Vertlib, V., 1978. Report on results of group geological mapping, scale 1:200,000, at South Gobi. Middle Gobi, East Gobi Aimags of Mongolia in 1974-1977: Ministry of Geology file number 2724.

Ivanhoe Mines Press Release, January 27, 2012. Ivanhoe Mines Expects to Maintain a Majority of Independent Directors on the Ivanhoe Board. Retrieved from: http://www.turquoisehill.com/s/news_releases.asp?ReportID=504110.

Jamsrandorj, G., Diatchkov, S.A., 1996. Placer Deposits of Mongolia. SEG Newsletter, pp. 1 and 10–14, 24.

Jargalsaihan, D., Kazmer, M., Baras, Z., Sanjaadorj, D., 1996. Guide to the Geology and Mineral Resources of Mongolia. Ulaanbaatar, Geological Exploration, Consulting and Services (GCS) Co., Ltd, 221 p.

Kynisky, J., Chakhmouradian, A., Xu, C., Krhmicek, L., Galiova, M., 2011. Zirconium mineralization in peralkaline granites and associated pegmatites of the Khan Bogd Complex, southern Mongolia. The Canadian Mineralogist 49, 947–965.

Marinov, N.A., Hasin, R.A., Borzakovskiy, Y.A., Zonenshain, L.P. (Eds.), 1973. Geology of Mongolian People Republic, v.2. Moscow, Nedra, (in Russian).

Metal Mining Agency of Japan, 1997. Non-ferrous Metal Deposits of C.I.S. Countries and Mongolia. Metal Mining Agency of Japan, 20 p.

Mayne, S., 2006. Clearing Out the BHP Old Guard. Crikey. Retrieved from: http://www.crikey.com.au/2006/04/28/clearing-out-the-bhp-old-guard/?wpmp_switcher=mobile.

Mossakovsky, A.A., Tomurtogoo, O., 1976. Upper Paleozoic of Mongolia. M. Nauka, 127p.

Munkhtsengel, B., Ohara, M., Gerel, O., Dandar, S., Tsuchiya, N., 2006. Preliminary study of formation mechanism of the erdenetiin ovoo porphyry copper-molybdenum deposit and environmental effects of erdenet mine, Northern Mongolia. In: Proceedings of the 3rd International Workshop on Water Dynamics. In Water Dynamics by American Institute of Physics, pp. 2004–2007.

Nazarov, Yu, P., 2015. Copper super concentrate production technology for operating copper-molybdenum processing plants. Eurasian Mining 1, 25–30.

Sanjdorj, S., 2011. Memories of the exploration of the Oyu Tolgoi project. Ulaanbaatar, 48.

Sillitoe, R.H., Gerel, O., Dejidma, G., Gotovsuren, A., Sanjaadorj, D., Baasandorj, S., Dashiin, B.E., 1996. Mongolia's Gold Potential. Mining Magazine, pp. 12–15.

Turquoise Hill Press Release, September 22, 2003a. Significant New Style of Bornite-Rich Porphyry Discovered in Extension of Far North Zone in Mongolia as Drilling Encounters Richest Mineralization to Date. Far North Zone Renamed the 'Hugo Dummett Deposit' in Honour of Ivanhoe Executive's Exploration Work. Retrieved from: http://www.turquoisehill.com/s/news_releases.asp?ReportID=375353.

Turquoise Hill Press Release, November 3, 2003b. Ivanhoe buys 2% Royalty that BHP Billiton Had Retained in Ivanhoe's Copper-gold Discoveries in Mongolia. Retrieved from: http://www.turquoisehill.com/s/news_releases.asp?ReportID=375358.

Turquoise Hill Press Release, November 10, 2003c. New Independent Resource Estimate Increases Size of Hugo Dummett Deposit in Mongolia's South Gobi to 1.36 Billion Tonnes at 1.04% Copper and 0.15 g/t Gold. Retrieved from: http://www.turquoisehill.com/s/news_releases.asp?ReportID=375361.

USGS. Mineral Resources Online Spatial Data. Silver Bell Porphyry Copper Deposit in Arizona, United States. https://mrdata.usgs.gov/sir20105090z/show-sir20105090z.php?id=79.

Watanabe, Y., Stein, H.J., 2000. Re-Os ages for the Erdenet and Tsagaan Suvarga porphyry Cu-Mo deposits, Mongolia, and tectonic implications. Economic Geology 95, 1537–1542.

Wainwright, A.J., 2008. Volcano Stratigraphic Framework and Magmatic Evolution of the Oyu Tolgoi Porphyry Cu-au District, South Mongolia. Dissertation of Doctor of Philosophy at the University of British Columbia, Vancouver, p. 277.

Wainwright, A., Tosdal, R., Lewis, P., Friedman, R., 2017. Exhumation and preservation of porphyry Cu-Au deposits at Oyu Tolgoi, South Gobi Region, Mongolia. Economic Geology 112, 591–601.

York, G., 2005. Desert Storm. Globe and Mail. Published Friday Sept. 30, 2005, Last Updated Tuesday, Mar. 17, 2007. Retrieved from: http://www.theglobeandmail.com/report-on-business/rob-magazine/desert-storm/article18249363/?page=all.

Glossary

A

absolute age
accretion
aegirine
age measurement
Agency for Foreign Investment and Foreign Trade
aggregate
aimak
Airag Uul
airborne gravity system
Ajo
Alaska
alaskite
albitization
albite
alkaline
Altai
Altaids
alteration
alumina
aluminate
alunite
alunitization
AMEC Company
Amphibolite
Anaconda
andesite
andesitic
andesite-dacitic
Andean type
Anglo Gold
Anglo American
anomaly
anticline
Antofagasta Minerals
Apatite
Apache Group
aphanitic
aplite
aquifer
arfvedsonite
argillization
arsenic
asthenosphere
ASARCO
Australia
A-vein
azurite

B

background level
barium
Bayankhongor zone
Bayannur
Bayan Bulag
Bayan Ovoo
Bayan Zurkh
bedrock
Beenup
beneficiation complex
Berkh
Berkley
BHP
BHP Billiton Ltd.
BHP Billiton merger
BHP Billiton PLC
BHP Branch
BHP Minerals
BHP Minerals XXK
Billiton
biotite
biotitization
bismuth
Blackwater
BLEG
BLEG sampling
Boeing
Bogd
Boodarie
bornite
Boroo
Bortolgoi
Botswana
Bowen coal basin
breccias
brecciation
Burma
Burvod story
B-vein

C

Cadaster Office
calcareous
California
Cambrian
Can-Asia
cap-rock
Carboniferous
Carlin type
Carlin style
Cascadia
Central Oyu
chalcocite
chalcocite blanket
chalcopyrite
chargeability
Chemex Laboratory
Chevron
Chief Executive
Chief Geologist
Chlorite
chloritization
Choir
chrysocolla
chromium
Chuquicamata
classification of mineral deposits
clastic sedimentary
claw-back terms and conditions
coal-bed
coalfield
coking coal
collision
Colorado University
concentrate
conglomerate
copper and gold concentrate
copper occurrence
copper porphyry
copper prospect
core logging
Cordillera Roundup
covellite
Cretaceous

D

dacite
Darkhan
deep drilling
deep penetrating IP
Delgerekh
Democratic Union
Devonian
diabase
diamond core drilling
Diavik
dickite
digenite
diorite
dickite-pyrophyllite
Discovery Group
Discovery Unit
discovery culture
discovery team
dinosaur
Dornod
Dripping Springs formation
drilling program
Duke Energy
D-vein
Dzhezkazgan

E

earn-in agreement
earn-in period
Ekati
Energy Fuels
England
enrichment
Entrée Gold
epidote
Erdenet
Erdenet-Magma JV
Erdenet-Magma Management Committee
Erdenetiin Ovoo
Escabrosa formation
Escondida
Eurasian region
exploration license fee
exploration license transfer
exploration model
external exploration funding

F

Falcon system
Far North Oyu
favorable environment
Feasibility Study
Feldspathic
first-class deposit
first right of refusal
fluorite
fly-in-fly-out
Foreign Investment Law
Freeport

G

gabbro
gamma spectrometry
garnet
geochemical
geochemistry
Geological Archive
geophysical
geoscientist
geosyncline
gher
Globe-Miami
Glukhariniy
Gneiss
Gobi
Gobi Geo
Gobi Altai
Gobi Desert
Gobi Drilling
gold export tax
golden triangle
Goonyella
granite
granitoid
granodiorite
Grasberg
Great Khural
Gunii Hooloi
Gurvan Saikhan

H

Hartley
Har Tolgoi
HBI (hot briquetted iron)
heap leaching
Helecho
hematite
high-grade core
high-grade hypogene copper-gold mineralization
Hong Kong
Honorary Vice-President
hornblende
Hugo Dummett Deposit
Hugo Mine
hydrothermal
hypogene mineralization

I

Ih Shankh
illite
induced polarization
Inner Mongolia
in-situ leaching
integrated approach
interpretation
intrusion or intrusive
IP anomaly
IP chargeability
Irkutsk Polytechnic Institute
island arc
isoclinal
Ivanhoe Mines
Ivanhoe Mines and Rio Tinto Agreement

J

jarosite
jasperoid
Java Gold
Javkhalant
Jiauquan Steel Company
joint-venture
JORC
Jubilee
Jwaneng

K

Kalamazoo
kaolin
kaolinitization
Kazakhstan
Kennecott
Khairkhan
Khanbogd complex
Khentei
Khokh Khad
Khovd Geological Expedition
Khubsugul Geological Expedition
Khuh Adar
Khurmen
Kilborn Engineering Pacific Ltd.
Kuala Lumpur
Kucing Liar
Kuroko type

L

Labor Law
Lac de Gras
lacustrine
Lake Khubsugul
Laramide
lead
legislature
Leica

limonite
limonitization
L'viv State University

M

Magma Copper Company
Magma porphyry deposit
magmatic
magnetite
magnetization
magnetic susceptibility
magnetic survey
mafic mineral
malachite
Maliy At-Uriah
Manakht
manganese
mantle plume
marble
massive
massive sulphide
Mesozoic
metamorphosed
metarhyolite
metasediment
metavolcanic
mineral occurrence
Mineral Resources Authority of Mongolia
mineral potential
mineralization
mineralogist
Minorco
molasse
molybdenite
monzonite
MonMap Company
Monywa
morphology
Moscow Geological Institute
Mount Whaleback
mudstones
Muruntau deposit
Muruntau type
Museum of Paleontology
Myanmar

N

Naadam
Nacho Formation
Nariin Zagiin Hooloi
National University of Mongolia
native copper
net smelter royalty
Nevada
Newcastle
Newmont Mining Corporation
NI 43-101
niobium
nonperforming asset
Noranda
North America
North-East Asia
North Oyu
Norwest Corporation
Northwest Territories
Novosibirsk State University
Nuurin

O

occurrence
October
Ok Tedi
Olympic Dam
Orapa
Orkhon-Selenge zone
Orogenic
Ovoot
oxidation
oxide-reduction
Oyu Tolgoi
Oyu Tolgoi license reduction
Oyu Tolgoi resource expansion oxidation

P

Pakistan
paleo environment
Paleozoic
Panama
PDAC international convention
PDAC Committee
PDAC core display
Pebble
Pennsylvanian
People Revolutionary Party
Peoples Friendship University
People's Republic of China
Permian
Peru
Phelps Dodge
phenocryst
Pinto Valley
phyllic

phyllite
PICT culture
PIMA
Pinal schist
plagioclase
plagiogranite
platform
Pliocene-Pleistocene
pluton
points of mineralization
polarization
porphyry gold-copper deposit
porphyry system
Port Hedland
post-mineral
potassic
potassium feldspar
prefeasibility
prioritization
primary covellite
primary mineralization
production growth
Puerto Rico
Pulkovo system
pyrite
pyrite-chalcopyrite mineralization
pyrite-covellite association
pyritization
pyrophyllite
pyroxene
pyrophylitization

Q

quartz-bornite-magnetite stockwork
quartz stockwork
quartz vein
quartz-magnetite
quartz-pyrite
quartzite
Quaternary
Queensland

R

ranking
regional geology
reconnaissance
red-bed
relic sulphide
Regional Chief Geophysicist
Reko Diq
resource
reserve
resistivity
Resolution
resource calculation
resource modelling
retrograde metamorphism
reverse circulation drilling
Rio Tinto
Riphean
Robinson
Rosco Postle Associates
royalty
Russia
Russian Federation
rutile

S

Sainshand
San Francisco
San Manuel
sandstone
schist
schistosity
secondary sulfide enrichment
sedimentary
Selenga
Senior Vice-President
sericite
sericitic
Shell Oil
Shirt
Shivee Oboo
Shuteen
Siberia
Siberian platform
silicification
Silurian
Silver Bell
Skarn
smektite
Society of Economic Geologists SEG
sole exploration funding
South America
South Gobi
South Oyu
Southwest Oyu
Soviet Union
sphalerite
SPOT
stability agreement
Standford University
step-by-step approach
strata-bound

strategic commodity
strategy
stratum
stockwork
subduction
Sudbury
Sukhoi Log
sulfide
supergene
supergene enrichment
supergene mineralization
Superior Mine
Superior Oil
Sweden
SX-EW solvent extraction & electro winning
syenite
syenodiorite
syncline

T

Table Mountain
Talnakh
Tampa speech
Tavan Tolgoi coal deposit
tectonic
Tekhnoexport
telescoping
tenorite
terrain
tetrahedrite
texture
thermal coal
thrust
Tianjin
Tier 1 deposit
Tier 2 deposit
tin
Tintic
Tintaya
titanite
tonalite
transient electromagnetics (TEM)
Trans-Mongolian railway
Triassic
Troy Resources
Tsagaan Suvarga
Tsankhi
Tsogtsetsii
tungsten
turquoise
Turquoise Hill
Tuva

U

UAZ truck
Udachnaya
Ukraine
Ukhaa Khudag
Ulaanbaatar
Ulaan Uul
ultra-alkaline
Ulziit
Ulyastai
University of Arizona
University of California
University of Chile
University of Ontario
University of Sydney
USGS
USSR
Utah

V

Vendian
vein
veinlet
vitrinite
VMS
Vodorazdelny
volcanic
volcanogenic

W

Washington State University
Western Mining Corporation
WMC
work commitment
World Bank

Y

Yandi

Z

Zambain Khudag
Zamtiin khad
Zavkhan
Zimbabwe
zinc
zirconium
Zost Uul
zunyite

Index

'*Note*: Page numbers followed by "f" indicate figures and "t" indicate tables.'

D

E

S

T

U

V

W

Z

Kolibri 360 GmbH
Ferdinand-Jühlke-Straße 7
99095 Erfurt, Deutschland
produktsicherheit@kolibri360.de